Exam Secrets

AS

Chemistry

John Sadler
Rob Ritchie

Contents

Examination boards

AQA www.aqa.org.uk	**CCEA** www.ccea.org.uk	**EDEXCEL** www.edexcel.org.uk	**OCR** www.ocr.org.uk	**WJEC** www.wjec.co.uk
5421	1110	8080	3882	330 80
Unit 1 1 hour 60 marks 30% Structured questions Atomic Structure, Bonding, Periodicity	Assessment Unit AS1 $1\frac{1}{2}$ hours 80 marks 35% Multiple choice and structured questions Atomic structure, Bonding, Shapes of molecules, Atomic size, Intermolecular forces, Energetics, Redox, The Periodic Table, Group 7, Titrations	Unit 1 1 hour 60 marks 30% Structured and extended questions Atomic structure, Formulae, Equations, Moles, Structure, Bonding, Periodic Table, Introduction to oxidation and reduction, Groups 1, 2 and 7	Unit 2811 1 hour 60 marks 30% Structured and extended questions Atoms, molecules, stoichiometry; Atomic structure; Chemical bonding and structure; Introduction to the Periodic Table; Groups 2 and 7	Unit CH1 $1\frac{1}{2}$ hours 66 marks 35% Structured and multiple choice questions Atomic structure, The Mole and stoichiometry, Structure and bonding, Gases, liquids and solids, The Periodic Table.
Unit 2 1 hour 60 marks 30% Structured questions Energetics, Kinetics, Equilibria, Redox, Group VII, Extraction of metals	Assessment Unit AS2 $1\frac{1}{2}$ hours 80 marks 35% Multiple choice and structured questions Isomerism of organic compounds, Hydrocarbons, Alkanes, Alkenes, Halogenoalkanes, Alcohols, Equilibrium, Kinetics, Group 2, Identification tests.	Unit 2 1 hour 60 marks 30% Structured questions Energetics, Alkanes, Alkenes, Halogenoalkanes, Alcohols, Equilibria, Industrial inorganic chemistry	Unit 2812 1 hour 60 marks 30% Structured and extended questions Chains and rings (basic concepts); Alkanes; Hydrocarbons (fuels, alkenes); Alcohols; Halogenoalkanes.	Unit CH2 $1\frac{1}{2}$ hours 66 marks 35% Structured and multiple choice questions Principle of energetics, Organic compounds: nomenclature, isomerism, reaction types, functional groups, Hydrocarbons and petroleum, Principles of chemical and acid-base equilibria, Chemical kinetics, Industrial and environmental aspects.
Unit 3a 1 hour 60 marks 25% Structured and extended questions Nomenclature, Isomerism, Petroleum, Alkanes, Epoxyethane, Halogenoalkanes, Alcohols	Assessment Unit AS3 $2\frac{1}{2}$ hours 65 marks 30% Planning exercise (20 marks) Structured question (20 marks) Two practical exercises (25 marks)	Unit 3A *either* $1\frac{3}{4}$ hours 50 marks 20% Internal assessment of practical skills *or* Internal assessment	Unit 2813/1 $\frac{3}{4}$ hour 45 marks 20% Structured and extended questions Enthalpy changes; Reaction rates; Chemical equilibrium	Unit CH3a $\frac{3}{4}$ hour 30 marks 10% Structured questions Theory-Experimental Interface Paper AND
Unit 3b 30 marks 15% *either* Centre Assessed Coursework *or* 2 hours 30 marks 15% Practical Examination		Unit 3B 1 hour 50 marks 20% Structured questions Laboratory chemistry	*either* 2813/2 Coursework 20% *or* 2813/3 $1\frac{1}{2}$ hours 20% Practical examination	*either* Unit CH3b 103 marks 10% Internally marked – externally moderated *or* Unit CH3c 103 marks 10% Externally marked and moderated

Nuffield 8086	**Salter's** 3887
Unit 1 $1\frac{1}{4}$ hours 60 marks 30% Structured questions Metal compounds: an introduction to inorganic chemistry; Alcohols: an introduction to organic chemistry; The Periodic Table and atomic structure; Acid-base reactions and the alkaline earth elements; Energy changes and reactions.	Unit 2850 $1\frac{1}{4}$ hours 75 marks 30% Structured and extended questions Elements of life; developing fuels
Unit 2 $1\frac{1}{2}$ hours 60 marks 40% Structured and extended questions Redox reactions and the halogens; Covalency and bond breaking; Organic chemistry: Hydrocarbons; Intermolecular forces; Organic chemistry: Halogenoalkanes.	Unit 2848 $1\frac{1}{2}$ hours 90 marks 40% Structured and extended questions From minerals to elements; The atmosphere; The polymer revolution
Unit 3 60 marks 30% Internal assessment of practical skills	Unit 2852/01 2 weeks 45 marks 15% Open-book paper Unit 2852/02 45 marks 15% Experimental skills

AS exams

Different types of questions

We have used structured questions in this exam practice book.

Structured questions

The main type of question you are likely to meet in a unit assessment is the structured question. These questions consist of an introduction, sometimes with data, followed by three to six parts, each of which may be further sub-divided. The introductory data provides the major part of the information to be used, and indicates clearly what the question is about.

- Make sure you read and understand the introduction before you tackle the question.
- Keep referring back to the introduction for clues to the answers to the questions.

Structured questions usually start with easy parts and get harder as you go through the question. Also you do not have to get each part right before you tackle the next question.

What examiners look for

- Examiners are obviously looking for the right answer, however it does not have to match the wording in the examiner's marking scheme exactly.
- Your answer will be marked correct if it contains all the main facts. You do not get extra marks for writing a lot of words.
- You should make sure that your answer is clear, easy to read and concise.
- You must make sure that your diagrams are neatly drawn with a ruler and labelled.

What makes an A, C and E grade candidate?

Obviously, you want to get the highest grade that you possibly can. The way to do this is to make sure that you have a good all round *knowledge* and *understanding* of Chemistry.

A grade candidates have a wide knowledge of chemistry and can apply that knowledge to novel situations. They are equally strong on all of the modules. The minimum mark for an A grade candidate is 80%.

C grade candidates have a reasonable knowledge of chemistry, but they are unsure about applying their knowledge to novel questions. They have weaknesses in some of the modules. The minimum mark for a C grade candidate is 60%.

E grade candidates have a limited knowledge of chemistry and have not learnt to apply their ideas to novel situations. They find it hard to memorise definitions and facts. The minimum mark for an E grade candidate is 40%.

Successful revision

Revision skills

- Try to do your revision in the same place and at about the same time every day.
- Always start with a topic with which you are familiar.
- Stop before you get too tired.
- Leave something easy with which to start your revision the next day.
- Try using post-it notes to keep track of your revision and highlighter pens to emphasise important points in your notes.
- Soothing background music can help.
- Don't stay up late the night before an exam trying to learn new topics. You will have forgotten most of it by the morning and the lack of sleep will probably affect your performance in the exam.

Practice questions

This book is designed to help you get better results.

- Look at the grade A and C candidates' answers and see if you could have done better.
- Try the exam practice questions and then look at the answers.
- Make sure you understand why the answers given are correct.
- When you feel ready, try the AS mock exam papers.

If you perform well on the questions in this book you should do well in the examination. Remember that success in examinations is about hard work, not luck. Thomas Edison said that genius was 1% inspiration and 99% perspiration.

Planning and timing your answers in the exam

- You should spend the first few minutes reading through the whole question paper.
- Answer the question you think you can do best first. The easiest question is usually the first question on the examination paper.
- There is no need to write out the question, it wastes time and space.
- Use the mark allocation to guide you on how much to write. The number of lines is usually the number of marks for the question +1, thus a question worth 4 marks has 5 lines.
- You should aim to use 1 minute for each mark; thus if a question has 5 marks, it should take you 5 minutes to answer the question.
- Plan your answers; do not write down the first thing that comes into your head. Planning is absolutely necessary for free-response/extended answer questions.
- Try and allow time to read through your answers. Do not cross anything out at this stage unless you are changing the answer you have given. Examiners can only mark what you have written down. Any answer is better than no answer.

How to boost your grade

- Learn the definitions – these are easy marks in exams and they reward effort and good preparation. If you want to boost your grade, you **cannot** afford to miss out on these marks – they are very easy to get.
- Always give the full definitions, remember that in many definitions temperature and pressure are important, and so state them.
- Always include state symbols and units, particularly in questions on thermochemistry and electrochemistry, *(s) for solid, (l) for liquid, (g) for gas and (aq) for a solution in water.*
- Always write balanced equations including state symbols. This is one of the ways chemists communicate information. Make sure you use the right symbols in your equations. *For reversible reactions* $\leftrightarrows$, *equilibrium reactions* $\rightleftharpoons$ *and reactions that go in one direction only* $\rightarrow$.
- Make sure that the equation you have written does take place and balances. Ionic equations should be used where appropriate. In some equations, particularly those in organic chemistry, you can use [H] to represent a reducing agent and [O] to represent an oxidising agent.
- Learn a set method for solving a calculation and use that method. You will then be in a far stronger position for tackling any numerical problems.
- Check any calculations you have made at least twice, and make sure that your answer is sensible.
- Make sure that you are familiar with all the functions on your calculator and that you know how to key in positive and negative indices.
- Learn to interpret the answers displayed by your calculator: *a calculator display of* 4.2^{-02} *means* 4.2×10^{-2} *or 0.042.*
- Never borrow a calculator for an examination. Often keys need to be pressed in different orders with different makes and types of calculator. Borrowing will lead to confusion and mistakes.
- Make sure you know the difference between number of decimal places and number of significant figures. *25.696 to one decimal place is 25.7 and to 4 significant figures it is 25.70.*
- For numerical calculations, always include units.
- Always show the sign for enthalpy changes and oxidation numbers: e.g. *calcium has an oxidation number of* $+2$.
- Do not give ions as names of reagents. $Cr_2O_7^{2-}$ is NOT a reagent. *The reagent is potassium dichromate(VI).*
- Draw diagrams with a ruler or some other aid. Label with straight lines. Make sure that the apparatus you have drawn is safe (e.g. that it is not completely sealed) and that the apparatus you have drawn is 'real' apparatus.
- Give complete colour changes. The test for an alkene is **NOT** that bromine turns colourless, but *the colour change is from brown to colourless.*
- Write chemical names legibly. *If you spell ethanal as ethanæl it will be marked wrong.*
- Read the question twice and underline or highlight key words in the questions. e.g. *in* ***terms of covalent bonding*** *suggest why nitrogen is* ***less reactive*** *than chlorine.*
- Remember that conditions of a reaction are very important, e.g. *whether a concentrated or a dilute acid is required*. Heating is not the same as heating under reflux which involves the use of a condenser.
- Use the Periodic Table and other data that you are given. *Don't try to remember data such as relative atomic masses.*
- You must represent the formulae of organic compounds correctly, so make sure you know the difference between empirical, molecular, structural, displayed and skeletal formula.
- Your graphs must be correctly labelled, have a suitable scale so that it fills the graph paper and that the best line is drawn through the points. When plotting graphs, make sure that you include fully labelled axes and units.

Periodic Table

The Periodic Table

Key: relative atomic mass 1.0; atomic symbol H; atomic number 1; name Hydrogen

1	2											3	4	5	6	7	0
																	4.0 He 2 Helium
6.9 Li 3 Lithium	9.0 Be 4 Beryllium											10.8 B 5 Boron	12.0 C 6 Carbon	14.0 N 7 Nitrogen	16.0 O 8 Oxygen	19.0 F 9 Fluorine	20.2 Ne 10 Neon
23.0 Na 11 Sodium	24.3 Mg 12 Magnesium											27.0 Al 13 Aluminium	28.1 Si 14 Silicon	31.0 P 15 Phosphorus	32.1 S 16 Sulphur	35.5 Cl 17 Chlorine	39.9 Ar 18 Argon
39.1 K 19 Potassium	40.1 Ca 20 Calcium	45.0 Sc 21 Scandium	47.9 Ti 22 Titanium	50.9 V 23 Vanadium	52.0 Cr 24 Chromium	54.9 Mn 25 Manganese	55.8 Fe 26 Iron	58.9 Co 27 Cobalt	58.7 Ni 28 Nickel	63.5 Cu 29 Copper	65.4 Zn 30 Zinc	69.7 Ga 31 Gallium	72.6 Ge 32 Germanium	74.9 As 33 Arsenic	79.0 Se 34 Selenium	79.9 Br 35 Bromine	83.8 Kr 36 Krypton
85.5 Rb 37 Rubidium	87.6 Sr 38 Strontium	88.9 Y 39 Yttrium	91.2 Zr 40 Zirconium	92.9 Nb 41 Niobium	95.9 Mo 42 Molybdenum	– Tc 43 Technetium	101 Ru 44 Ruthenium	103 Rh 45 Rhodium	106 Pd 46 Palladium	108 Ag 47 Silver	112 Cd 48 Cadmium	115 In 49 Indium	119 Sn 50 Tin	122 Sb 51 Antimony	128 Te 52 Tellurium	127 I 53 Iodine	131 Xe 54 Xenon
133 Cs 55 Caesium	137 Ba 56 Barium	139 La 57 Lanthanum	178 Hf 72 Hafnium	181 Ta 73 Tantalum	184 W 74 Tungsten	186 Re 75 Rhenium	190 Os 76 Osmium	192 Ir 77 Iridium	195 Pt 78 Platinum	197 Au 79 Gold	201 Hg 80 Mercury	204 Tl 81 Thallium	207 Pb 82 Lead	209 Bi 83 Bismuth	– Po 84 Polonium	– At 85 Astatine	– Rn 86 Radon
– Fr 87 Fracium	– Ra 88 Radium	– Ac 89 Actinium	– Rf 104 Rutherfordium	– Db 105 Dubnium	– Sg 106 Seaborgium	– Bh 107 Bohrium	– Hs 108 Hassium	– Mt 109 Meitnerium	– Unn 110 Ununnilium	– Uuu 111 Unununium	– Uub 112 Ununbium		– Uuq 114 Ununquadium		– Uuh 116 Ununhexium		– Uuo 118 Ununoctium

lanthanides	140 Ce 58 Cerium	141 Pr 59 Praseodymium	144 Nd 60 Neodymium	– Pm 61 Promethium	150 Sm 62 Samarium	152 Eu 63 Europium	157 Gd 64 Gadolinium	159 Tb 65 Terbium	163 Dy 66 Dysprosium	165 Ho 67 Holmium	167 Er 68 Erbium	169 Tm 69 Thulium	173 Yb 70 Ytterbium	175 Lu 71 Lutetium
actinides	– Th 90 Thorium	– Pa 91 Protactinium	– U 92 Uranium	– Np 93 Neptunium	– Pu 94 Plutonium	– Am 95 Americium	– Cm 96 Curium	– Bk 97 Berkelium	– Cf 98 Californium	– Es 99 Einsteinium	– Fm 100 Fermium	– Md 101 Mendelevium	– No 102 Nobelium	– Lr 103 Lawrencium

Chapter 1

Atomic structure

Questions with model answers

C grade candidate – mark scored 8/14

In terms of atomic structure, give **one** similarity and **one** difference between each of the following pairs of substances: [7 × 2]

(a) A proton and a neutron

Protons and neutrons both have the same mass, ✓
a proton has a charge of +1, a neutron has no charge. ✓

(b) A proton and a hydrogen atom

They both have one proton, ✓ the proton has a positive charge. ✗

(c) $^{12}_{6}C$ and $^{14}_{6}C$

They have the same number of protons (6), ✓ but a different number of neutrons (6 and 8). ✓

(d) O and O^{2-}

They have the same number of protons (8), ✓ O^{2-} has two extra electrons. ✓

(e) Cl^- and Cl^+

They have the same number of protons, ✓ Cl^+ has more electrons. ✗

(f) Na^+ and F^-

They are both ions ✗ but have different charges. ✗

(g) $^{55}_{25}Mn^{2+}$ and $^{56}_{26}Fe^{3+}$

They are both transition metals ✗ but have different charges. ✗

Examiner's Commentary

For help see Revise AS Study Guide 1.1

(b) Hydrogen has one electron, a proton has no electrons.

(e) Cl^- has 18 electrons, Cl^+ has 16 electrons.

(f) True, but this is NOT the answer in terms of atomic structure. Both have the same number of electrons (10), but a different number of protons.

(g) Again true, but you have not answered the question. They have the same number of neutrons and electrons but a different number of protons.

GRADE BOOSTER

This is an example of where the candidate has not read the question carefully. Although the chemistry is correct, the answers are not for the questions asked.

A grade candidate – mark scored 8/11

Examiner's Commentary

(a) Electrons have very small masses and they are negatively charged. Give one other property of electrons. [1]

they spin ✓

Electrons spin on their axes, clockwise and anticlockwise. These are represented by ↑ and ↓.

(b) Which noble gas has no 'p' orbitals? [1]

helium ✓

(c) Listed below are the electronic configurations of 5 elements, R to V. One of the configurations is wrong. [9]

R $1s^2\ 2s^2\ 2p^6$

S $1s^2\ 2s^2\ 2p^6\ 3s^1$

T $1s^2\ 2s^2\ 2p^6\ 3s^2\ 3p^6\ 3d^5\ 4s^1$

U $1s^2\ 2s^2\ 2p^4\ 3s^1$

V $1s^2\ 2s^1$

(i) Identify the element with the wrong electronic configuration and, assuming that the number of electrons are correct, write down its correct electronic configuration.

U: $1s^2\ 2s^2\ 2p^5$ ✓

(ii) Which element is a noble gas?

R ✓

(iii) Which **two** elements are in the same group of the Periodic Table?

S and V ✓

Elements in the same group have the same number of electrons in their outermost shell.

(iv) Identify element T.

Cr ✓

(v) Why is the 4s sub-shell filled before the 3d sub-shell?

The 4s sub-shell is in a lower energy state than the 3d sub-shell. ✓

(vi) Suggest why T has the structure shown and not $1s^2\ 2s^2\ 2p^6\ 3s^2\ 3p^6\ 3d^4\ 4s^2$.

The 4s sub-shell is in a lower energy state than the 3d sub-shell. ✗

Half-filled shells are more stable.

(vii) Divide the elements R to V into s block, p block and d block elements in the Periodic Table.

s block: S,U,V; ✗ p block: R; ✗ d block: T ✓

Remember that U is incorrect and is in the p-block (see **(i)** above).

Atomic structure

Exam practice questions

1 The table gives the electronic configuration of consecutive elements in each of two groups (**A** and **B**) of the Periodic Table.

Group **A**		Group **B**	
Element	Configuration	Element	Configuration
T	$1s^2\ 2s^1$	X	$1s^2\ 2s^2\ 2p^4$
V	$1s^2\ 2s^2\ 2p^6\ 3s^1$	Y	
W		Z	$1s^2\ 2s^2\ 2p^6\ 3s^2\ 3p^6\ 3d^{10}\ 4s^2\ 4p^4$

(a) In which group of the Periodic Table are:

(i) T, V and W [1]

(ii) X, Y and Z? [1]

(b) Write down the electronic configuration of:

(i) W [1]

(ii) Y. [1]

(c) For the electron in $3s^1$, what is meant by 3, s and 1? [3]

(d) How many pairs of electrons are there in:

(i) V [1]

(ii) Z? [1]

(e) What is meant by an 'orbital'? [2]

(f) Element V reacts with element X to form a compound.

(i) What is the formula of the compound formed between V and X? [1]

(ii) What type of bonding would you expect in this compound? Give a reason for your answer. [2]

[Total: 14]

Atomic structure

Answers on pages 14–15 Answers on pages 14–15 Answers on pages 14–15

2 The diagram below shows the first ionisation energies of the elements in the third period of the Periodic Table.

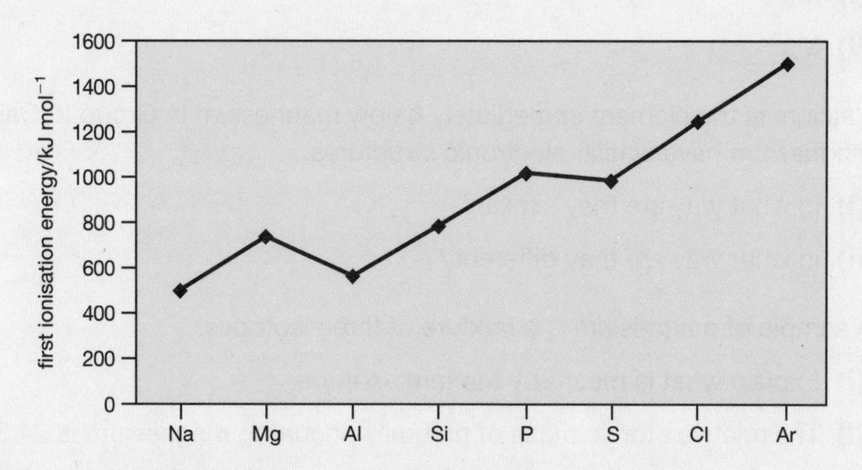

(a) The electronic configuration of silicon can be written as [Ne] $3s^2\ 3p_x^1\ 3p_y^1$

(i) In terms of electron configuration, what does [Ne] represent? [1]

(ii) Using the same format, write down the electronic configurations of Mg and S. [2]

(b) Explain, including an equation with state symbols, the meaning of *first ionisation energy*. [4]

(c) State **three** factors that affect ionisation energy. [3]

(d) Sketch, including approximate values, the first ionisation energies of the elements carbon to aluminium in the Periodic Table. Explain the shape of your sketch. [2]

[Total: 12]

Answers on pages 14–15 Answers on pages 14–15 Answers on pages 14–15

3 **(a)** Write the full electronic configuration in terms of sub-shells for:

(i) Mg [1]

(ii) Mg^{2+}. [1]

(b) Calcium is the element immediately below magnesium in Group 2. Calcium and magnesium have similar electronic structures.

(i) In what way are they similar? [1]

(ii) In what way are they different? [1]

(c) A sample of magnesium is a mixture of three isotopes.

(i) Explain what is meant by the term *isotope*. [2]

(ii) The relative atomic mass of naturally occurring magnesium is 24.32.

Use the information in the table below to calculate the relative isotopic mass of the third isotope of magnesium and then complete the table.

Isotope	Percentage abundance
$^{24}_{12}Mg$	79.0%
$^{25}_{12}Mg$	10.0%

[4]

(d) Calcium has a relative atomic mass of 40.1. It has four isotopes. Suggest, with a reason:

(i) the atomic number [1]

(ii) the mass number of the most abundant isotope of calcium. [1]

[Total: 12]

Atomic structure

Answers on pages 14–15 Answers on pages 14–15 Answers on pages 14–15

4 The table below shows seven successive ionisation energies for the first four elements of a group, in order of increasing atomic mass. (The letters are **not** the symbols of the elements.)

Element	Ionisation energies/kJ mol^{-1}						
	1	2	3	4	5	6	7
V	1090	2400	4600	6200	37 800	47 000	no value
W	790	1600	3200	4400	16 100	20 000	23 600
X	760	1500	3300	4400	8950	11 900	14 900
Y	710	1400	2900	3900	7780	9770	1250

(a) Write an equation, including state symbols, for the *2nd ionisation energy* of element **X**. [2]

(b) Sketch a graph of the number of electrons removed from the atom **W** against ionisation energy. [2]

(c) State, with a reason, in which group of the Periodic Table you would find elements **V, W, X** and **Y**. [2]

(d) Explain why the value of the 1st ionisation energies decreases down the group. [2]

(e) **(i)** State, with a reason why the 7th ionisation energy of elements **W, X** and **Y** is larger than the 6th ionisation energy. [1]

(ii) Explain why element **V** does **not** have a value for the 7th ionisation energy. [1]

[Total: 10]

Atomic structure

Answers

(1) (a) (i) Group 1 ✓ **(ii)** Group 6 ✓

examiner's tip

To find the group, add up the electrons in the highest principal quantum number. In group A there is only 1 electron and in group B there are 6 electrons.

(b) (i) W is $1s^2\ 2s^2\ 2p^6\ 3s^2\ 3p^6\ 4s^1$ ✓ **(ii)** Y is $1s^2\ 2s^2\ 2p^6\ 3s^2\ 3p^4$ ✓

(c) 3 is the principal quantum number; ✓ s is the sub-shell; ✓ 1 is the number of electrons in the sub-shell. ✓

(d) (i) 5 ✓ **(ii)** 16 ✓

examiner's tip

Each orbital can hold 2 electrons. There is one s orbital that can hold 1 pair of electrons, three p orbitals that can hold 6 electrons (3 pairs); five d orbitals that hold 10 electrons (5 pairs) and seven f orbitals that hold 14 electrons (7 pairs). The orbitals in a sub-shell are occupied singly before pairing starts.

(e) A region around an atom where there is a high probability of finding an electron ✓ at any moment in time. ✓

(f) (i) V_2X ✓

(ii) Ionic; ✓ electrons transferred from V to X to obtain noble gas structures. ✓

---

(2) (a) (i) $1s^2\ 2s^2\ 2p^6$ ✓

(ii) Mg is $[Ne]3s^2$ ✓ and S is $[Ne]3s^2\ 3p_x^2\ 3p_y^1\ 3p_z^1$ ✓

(b) The first ionisation energy of an element is the energy required to remove **1 electron** ✓ from each atom in **1 mole** of **gaseous atoms** ✓ to form 1 mole of **gaseous 1^+ ions.** ✓ $E(g) \rightarrow E^+(g) + e^-$ ✓

examiner's tip

The key words are in bold.

(c) atomic radius; ✓ nuclear charge; ✓ electron shielding (screening) ✓

(d)

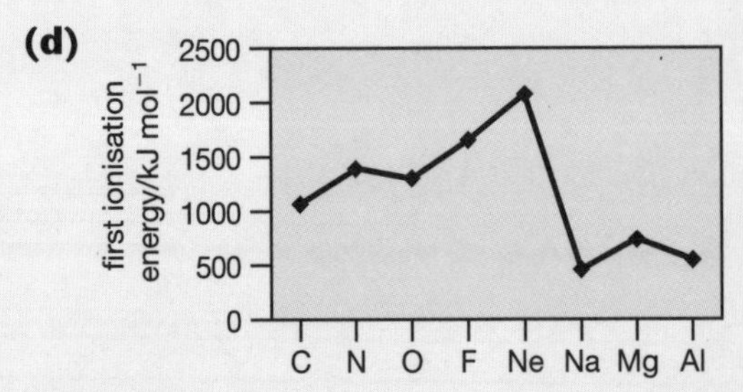

From C to Ne this is a similar shape to Si to Ar in the question, ✓ but values are higher because the electrons are nearer the nucleus. ✓

examiner's tip

Note, the last three elements in this graph were the first three elements in the question.

(3) (a) (i) $1s^2\ 2s^2\ 2p^6\ 3s^2$ ✓ **(ii)** $1s^2\ 2s^2\ 2p^6$ ✓

> **examiner's tip**
> **When electrons are removed from an atom, a positive ion is formed.**

(b) (i) Same number of electrons in the outermost shell (2 electrons) ✓

(ii) Calcium has one more shell ✓

> **examiner's tip**
> **Going down a group, the number of shells increases by one.**
> **Calcium is $1s^2\ 2s^2\ 2p^6\ 3s^2\ 3p^6\ 4s^2$**

(c) (i) Same number of protons ✓ different number of neutrons ✓ with different masses

(ii) % of third isotope will be 100 − 79 − 10 = 11% ✓

If n is the atomic mass of the third isotope then

$$\frac{(24 \times 79) + (25 \times 10) + (n \times 11)}{100} \qquad n = 26 ✓$$

Completion of table as ${}^{26}_{12}Mg$ ✓ and 11% ✓

> **examiner's tip**
> **Remember that isotopes of the same element have the same number of electrons. Each magnesium isotope will have 12 electrons.**

(d) (i) 20. For the first 20 elements, the number of protons is approximately equal to the number of neutrons in the nucleus. ✓

(ii) 40. Since the relative atomic mass is 40.1, the commonest isotope is likely to be 40 to ensure a value of 40.1. ✓

---

(4) (a) $X^+(g) \rightarrow X^{2+}(g) + e^-$ ✓ correct state symbols ✓

(b)

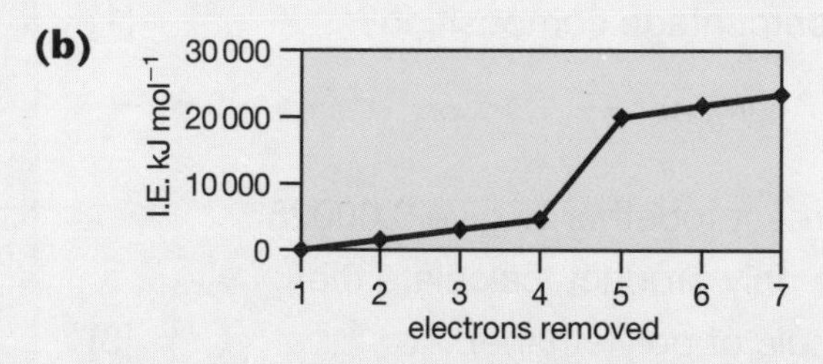

jump between 4 and 5 ✓
steady rise throughtout ✓

> **examiner's tip**
> **Make sure you have plotted the right element!**

(c) Group 4; ✓ there is a large jump in the value of I.E. between 4th I.E. and 5th I.E. ✓

(d) The greater the distance between the nucleus and the outer electrons, the less the attractive force. ✓

Electron shielding reduces the overall attractive force experienced by the outer electrons. ✓

> **examiner's tip**
> **Nuclear charge also affects the ionisation energy, but this effect is noticed going across a period, hence I.E. increases across a period.**

(e) (i) A larger amount of energy is required to remove the seventh electron from an ion with an overall charge of 6+. ✓

(ii) The element must only have 6 electrons. ✓

Chapter 2

Atoms, moles and equations

Questions with model answers

C grade candidate – mark scored 6/10

(a) Find **(i)** the empirical formula and **(ii)** the molecular formula, of a gaseous compound X containing 85.7% of carbon and 14.3% hydrogen. (M_r = 56. A_r : H, 1.0; C, 12.0) [4]

(i) Molar ratio of atoms $C = \frac{85.7}{12} = 7.14$;
and $H = \frac{14.3}{1} = 14.3$ ✓

Divide by smallest C : H = 1 : 2
Therefore the empirical formula is CH_2 ✓

(ii) Each CH_2 unit has a mass of 14 ✓
Number of units $= \frac{56}{14} = 4$
Molecular formula is C_4H_8 ✓

(b) Write the equation, including state symbols, for the complete combustion of X. [2]

$C_4H_8(g) + 6O_2(g) \rightarrow 4CO_2(g) + 4H_2O(l)$ ✓✓

(c) Compound X is an alkene. All alkenes have the formula C_nH_{2n}. Why do all alkenes have the same percentage composition? [1]

Because they contain the same number of carbon atoms. ✗

(d) If 7.0 g of ethene (C_2H_4) molecules react together to give 0.00025 moles of poly(ethene) $(C_2H_4)_n$ as the only product, calculate the number of carbon atoms in a molecule of poly(ethene). [3]

✗ ✗ ✗

Examiner's Commentary

Always divide by the A_r.

?
For help see Revise AS Study Guide 2.3

Complete combustion always gives CO_2 not CO.

The percentage will always be $\frac{12n}{14n} \times 100 = 85.7\%$
The M_r of $C_nH_{2n} = 12 \times n + 2 \times n = 14n$.

Number of moles of ethene $= \frac{7}{28} = 0.25$
(M_r of ethene $= 2 \times 12 + 4 \times 1 = 28$)
Since 0.00025 moles of polyethene is obtained from 0.25 moles of ethene then 1 mole of poly(ethene) will be obtained from $\frac{0.25}{0.00025}$ moles of ethene = 1000 moles.
Since each mole of ethene contains 2 carbon atoms, the number of carbon atoms in a molecule of poly(ethene) = 2000.

GRADE BOOSTER

Even if you cannot do all the question, you should make some attempt. Since the question was about moles, the candidate should have been able to work out the number of moles of ethene in 7.0 g of ethene.

A grade candidate – mark scored 8/10

(a) (i) What instrument is used to measure relative atomic mass? [1]

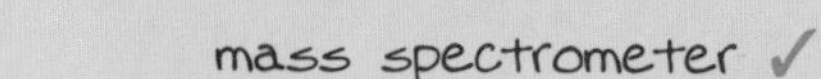

(ii) Naturally occurring magnesium consists of 78.6% ^{24}Mg, 10.1% ^{25}Mg and 11.3% ^{26}Mg.
Calculate to three significant figures, the relative atomic mass of naturally occurring magnesium. [2]

$$\frac{24 \times 78.6 + 25 \times 10.1 + 26 \times 11.3}{100}$$ ✓

$= 24.327$ ✗

(b) When magnesium burns in air, a mixture of magnesium oxide and magnesium nitride (Mg_3N_2) is formed. Magnesium nitride is a white solid. Write the equation, including state symbols, for magnesium reacting with nitrogen. [2]

$N_2(g) + 3Mg(s) \rightarrow Mg_3N_2(s)$ ✓✓

(c) Magnesium nitride reacts with water to give magnesium hydroxide and ammonia. The equation for this reaction is

$$Mg_3N_2(s) + 6H_2O(l) \rightarrow 3Mg(OH)_2(aq) + 2NH_3(g)$$

(i) How would you show that ammonia gas was given off? [1]

It turns damp litmus paper blue. ✓

(ii) What volume of ammonia (at r.t.p.) would be liberated if 2.018 g of magnesium nitride were added to excess water?
(A_r : Mg, 24.3; N, 14.0) [3]

1 mole of Mg_3N_2 gives 2 moles of ammonia (48 dm^3) ✓
Number of moles of Mg_3N_2 used is $\frac{2.018}{100.9} = 0.02$ moles ✓
Volume formed $= 48 \times 0.02 = 0.96\ dm^3$ ✓

(iii) Suggest why the volume of ammonia given off is much less than this value. [1]

The reaction did not go to completion. ✗

Examiner's Commentary

For help see Revise AS Study Guide 2.1

You have confused three significant figures with three decimal places (24.3).

Ammonia can always be recognised by its characteristic pungent smell.

One mole of gas at r.t.p. occupies 24 dm^3.

The water was in excess, so the reaction would have gone to completion. Ammonia is very soluble in water.

Exam practice questions

1 The mass spectrometer is used to measure the relative atomic mass of elements and the relative molecular mass of molecules. Its three main functions are to:

- ionise the sample
- separate the ions in the sample in terms of their mass-to-charge ratio
- collect and detect the ions and measure their relative abundance.

(a) **(i)** Will the sample be a gas, liquid or solid when it enters the mass spectrometer? [1]

(ii) How are the ions formed in the mass spectrometer?
Write an equation for the formation of M^+ from a sample M. [2]

(iii) After ionisation the ions are accelerated. How are the ions accelerated? [1]

(iv) Circle the ion which will be deflected the most
$^{20}Ne^+$, $^{21}Ne^+$, $^{22}Ne^+$, $^{20}Ne^{2+}$, $^{21}Ne^{2+}$ or $^{22}Ne^{2+}$.

Give a reason for your answer. [2]

(v) Why are very low pressures used in the ionisation chamber? [2]

(b) The diagram below shows the mass spectrum for naturally occurring gallium (Ga).

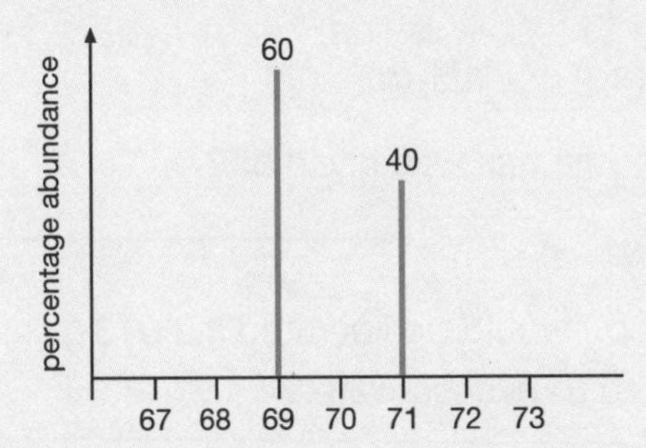

Calculate the relative atomic mass of naturally occurring gallium. [2]

(c) Bromine has two isotopes ^{79}Br and ^{81}Br. If Br_2 is used in the mass spectrometer there are three peaks of Br_2^+ at 158, 160 and 162.
Which species are responsible for these peaks? [1]

[Total: 11]

Answers on pages 22–23 Answers on pages 22–23 Answers on pages 22–23

2 One of the impurities present in fuels is hydrogen sulphide (H_2S). Hydrogen sulphide burns to form sulphur dioxide and water.

(a) 102.3 kg of hydrogen sulphide is completely burnt.

(i) Construct the equation for the combustion of hydrogen sulphide. [1]

(ii) How many moles of hydrogen sulphide have been burnt? [1]

(iii) What volume of sulphur dioxide (in dm^3) at r.t.p. have been formed? (A_r: H, 1.0; S, 32.1) [2]

(iv) How many molecules of sulphur dioxide have been formed? Avogadro's number $6 \times 10^{23}\ mol^{-1}$ [1]

(b) Sulphur dioxide pollution forms acid rain by reacting with oxygen in the air to form sulphur trioxide, which in turns reacts with water to form sulphuric acid.
Write balanced equations, with state symbols, for these two reactions. [3]

(c) One way of removing sulphur dioxide is to pass it through calcium hydroxide solution.

$$Ca(OH)_2(aq) + SO_2(g) \rightarrow CaSO_3(aq) + H_2O(l)$$

The solubility of calcium hydroxide in water is $1.85\ g\ dm^{-3}$.

(i) How many moles of calcium hydroxide dissolve in 1 dm^3 of water? (A_r: Ca, 40.1) [1]

(ii) What volume of sulphur dioxide can be absorbed by 100 dm^3 of this calcium hydroxide solution at r.t.p.? [2]

[Total: 11]

Answers on pages 22–23 Answers on pages 22–23 Answers on pages 22–23

3 This question is about an acid **X** and two gases **P** and **Q**.

(a) The relative molecular mass of acid **X** is 82. A student made up a solution of **X** in water with a concentration of 20.5 g dm^{-3}. In a titration, 10.0 cm^3 of this solution reacted exactly with 25.0 cm^3 of 0.200 mol dm^{-3} aqueous sodium hydroxide.

(i) Calculate the number of moles of **X** used in the titration. [2]

(ii) Calculate the number of moles of sodium hydroxide used in the titration. [1]

(iii) Calculate the number of moles of sodium hydroxide that reacts with 1 mole of acid **X**. [2]

(iv) Acid **X** has the molecular formula H_xSO_y. Deduce the values of *x* and *y* and hence the molecular formula of **X**. [3]

(v) Write the equation for the reaction between **X** and sodium hydroxide. [2]

(b) 1.00 dm^3 of gas **P** at room temperature and pressure (r.t.p.) has a mass of 1.25 g. The empirical formula of **P** is CH_2O.
1 mole of a gas at r.t.p. has a volume of 24.0 dm^3.

(i) What is meant by *empirical formula*? [1]

(ii) Calculate the relative molecular mass of **P**. [1]

(iii) Show that ethanoic acid has the same empirical formula as **P**. [2]

(c) Ethanoic acid boils at 118 °C. Some ethanoic acid was vaporised by heating it to just above its boiling point at atmospheric pressure. Under these conditions, 1 mole of ethanoic acid vapour has a volume of 32 dm^3.

What would be the mass of 1 dm^3 of ethanoic acid vapour under these conditions? [2]

[Total: 16]

Answers on pages 22–23 Answers on pages 22–23 Answers on pages 22–23

4 Tin forms two chlorides, tin(II) chloride $SnCl_2$ (boiling point 896 K) and tin(IV) chloride $SnCl_4$ (boiling point 386 K).

(a) Tin(IV) chloride can be prepared by reacting tin with excess chlorosulphonic acid, $ClSO_3H$, using the apparatus shown below.

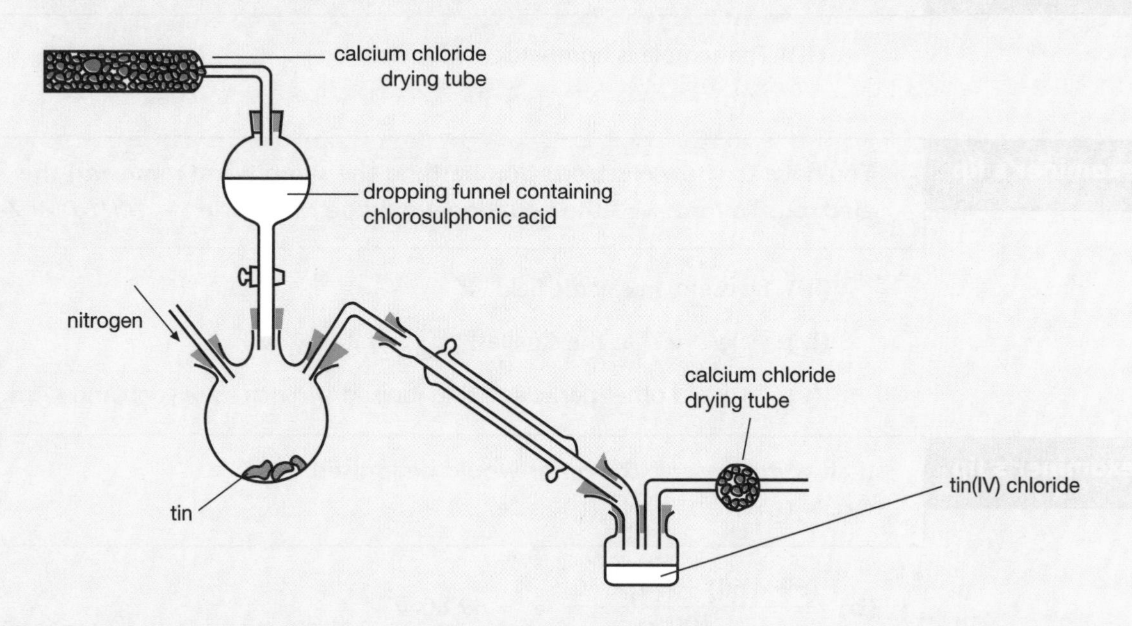

Look at the above diagram and identify **two** methods of preventing tin(IV) chloride and chlorosulphonic acid from coming into contact with water from the air. **[2]**

(b) The equation for the reaction is:

$$Sn(s) + 4ClSO_3H(l) \rightarrow SnCl_4(l) + 2SO_2(g) + 2H_2SO_4(l)$$

Name the gas given off during the above reaction. **[1]**

(c) What evidence is there that the reaction is exothermic (gives out heat)? **[1]**

(d) 23.8 g of tin were used in the reaction. What is the maximum mass of tin(IV) chloride that can be formed? **[3]**

(e) One mole of chlorosulphonic acid reacts with one mole of water to form one mole of acid **A** and one mole of acid **B**.

Write the equation for the reaction and name the acids **A** and **B**. **[3]**

[Total: 10]

Answers on pages 22–23 **Answers** on pages 22–23 **Answers** on pages 22–23

Answers

(1) (a) (i) gas ✓

examiner's tip **The sample must be in the vapour/gaseous state.**

(ii) The sample is bombarded with electrons ✓

$M(g) + e^- \rightarrow M^+(g) + 2e^-$ ✓

examiner's tip **You have to show electrons bombarding the sample and removing the electron. To form M^{2+} the equation would be $M^+(g) + e^- \rightarrow M^{2+}(g) + 2e^-$**

(iii) By using an electric field. ✓

(iv) $^{20}Ne^{2+}$ ✓ has the smallest $\frac{\text{mass}}{\text{charge}}$ ratio. ✓

(v) To avoid other particles being ionised ✓ such as oxygen and nitrogen. ✓

examiner's tip **If air were present, the gases would be ionised, e.g. $N_2(g) + e^- \rightarrow N_2^+(g) + 2e^-$**

(b) $\frac{(69 \times 60) + (71 \times 40)}{100}$ ✓ = 69.80 ✓

examiner's tip **Make sure that your answer is between the two isotopic masses.**

(c) $^{79}Br^{79}Br^+$; $^{79}Br^{81}Br^+$; $^{81}Br^{81}Br^+$ ✓

---

(2) (a) (i) $2H_2S(g) + 3O_2(g) \rightarrow 2H_2O(g) + 2SO_2(g)$ ✓

(ii) 102.3 kg = 102 300 g. Number of moles of $H_2S = \frac{102\,300}{34.1} = 3000$ ✓

(iii) 3000 moles of sulphur dioxide will be formed ✓

$= 3000 \times 24 = 72\,000$ dm^3 ✓

examiner's tip **From the equation 1 mole of hydrogen sulphide gives 1 mole of sulphur dioxide.**

(iv) number of SO_2 molecules $= 3000 \times (6 \times 10^{23}) = 1.8 \times 10^{27}$ ✓

(b) $2SO_2(g) + O_2(g) \rightarrow 2SO_3(g)$ ✓

$SO_3(g) + H_2O(l) \rightarrow H_2SO_4(aq)$ ✓ ✓ (state symbols in both equations)

(c) (i) Mass of $Ca(OH)_2 = \frac{1.85}{74.1} = 0.025$ moles ✓

(ii) Number of moles used $= 0.025 \times 100 = 2.5$ moles ✓

Volume of gas absorbed $= 2.5 \times 24 = 60$ dm^3 ✓

examiner's tip **You might be asked why it is more than this. The answer is that sulphur dioxide is soluble in water.**

(3) (a) (i) $\frac{20.5}{82}$ ✓ $\times \frac{10.0}{1000} = 0.0025$ moles ✓

(ii) moles $= \frac{0.200 \times 25.0}{1000} = 0.0050$ moles ✓

(iii) ratio $= 0.0025 : 0.0050$ ✓ $= 1 : 2$ ✓

examiner's tip

If the ratio does not come out to be a whole number ratio, you must have made a mistake in your calculations.

(iv) $x = 2$ ✓ $y = 3$ ✓ **X** is H_2SO_3 ✓

(v) $H_2SO_3(aq) + 2NaOH(aq) \rightarrow Na_2SO_3(aq) + 2H_2O(l)$ ✓ ✓ for balancing

(b) (i) Empirical formula represents the simplest, **whole number ratio of atoms** of **each element** in a compound. ✓

examiner's tip

The important words are in bold.

(ii) $1.25 \times 24 = 30$ ✓

(iii) molecular formula of ethanoic acid is $C_2H_4O_2$, ✓ simplest ratio of elements is CH_2O ✓

(c) Molar mass of $CH_3COOH = 60$ ✓

Mass of 1 $dm^3 = \frac{60}{32} = 1.875$ g ✓

---

(4) (a) Filling the apparatus with nitrogen ✓ and using calcium chloride drying tubes ✓

examiner's tip

When a question asks you to look at a piece of information it is guiding you to the correct answer. In this case, both answers were written on the diagram.

(b) sulphur dioxide ✓

(c) The tin is not heated, yet tin(IV) chloride comes off as a vapour. ✓

(d) moles of tin used $= \frac{23.8}{119} = 0.2$ moles ✓

moles of tin(IV) chloride formed = 0.2 moles; mass $= 0.2 \times 261$ ✓ $= 130.5$ g ✓

examiner's tip

1 mole of tin gives 1 mole of tin(IV) chloride. The relative molecular mass of $SnCl_4$ is 261.

(e) $ClSO_3H + H_2O \rightarrow H_2SO_4 + HCl$ ✓
sulphuric acid ✓ and hydrochloric acid ✓

Chapter 3

Structure and bonding

Questions with model answers

C grade candidate – mark scored 6/10

The atoms in molecules of chlorine and oxygen are held together by covalent bonds.

(a) What is meant by the term 'covalent bond'? [2]

A covalent bond is a shared ✓ pair of electrons. ✓

(b) Using 'dot and cross' diagrams, show the electronic arrangement in molecules of

(i) chlorine **(ii)** oxygen. [2]

Cl Cl ✓ O O ✗

(c) The diagram shows the covalent bonding in nitrogen.

N N

In terms of covalent bonding, suggest why nitrogen is less reactive than oxygen and chlorine. [1]

Because it has fewer electrons in its outermost shell. ✗

(d) Water is the hydride of oxygen. Explain why water has a higher melting point than the hydride of sulphur. [1]

Water has hydrogen bonding. ✓

(e) The hydride of nitrogen is ammonia.

(i) Draw a 'dot and cross' diagram to show the bonding in ammonia. [1]

H N H H ✓

(ii) Explain, with the aid of diagrams, the formation of the ammonium ion NH_4^+ from ammonia and a hydrogen ion H^+. [3]

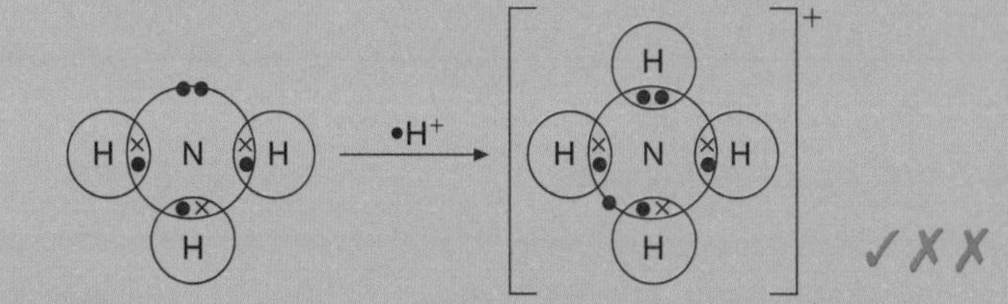

✓✗✗

Examiner's Commentary

Don't forget *pair* of electrons.

There must be 8 electrons in the outermost shell. You must distinguish between the electrons from each atom (dot and cross).

Triple covalent bonds are harder to break than single or double bonds.

Other inorganic compounds with hydrogen bonding are hydrogen fluoride and ammonia.

You have drawn a hydrogen atom and given the ammonium ion an extra electron. The hydrogen ion has no electrons and it forms a dative covalent bond with ammonia. (*See Study Guide 3.1.*)

GRADE BOOSTER

Always check that you have 8 electrons in the outermost shell.

A grade candidate – mark scored 13/16

Examiner's Commentary

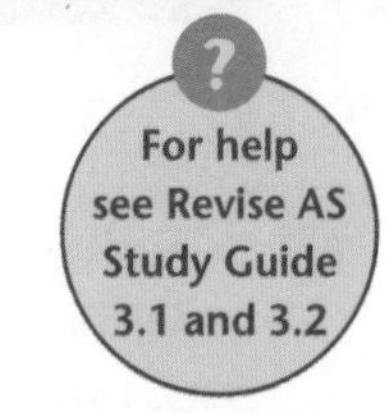

When hydrogen chloride is added to water, the following reaction occurs:

$H_2O(l) + HCl(g) \rightarrow H_3O^+(aq) + Cl^-(aq)$

(a) Draw the structure of **(i)** the Cl^- ion and **(ii)** the H_3O^+ ion. [2]

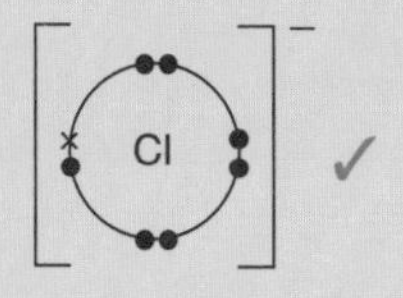

✓

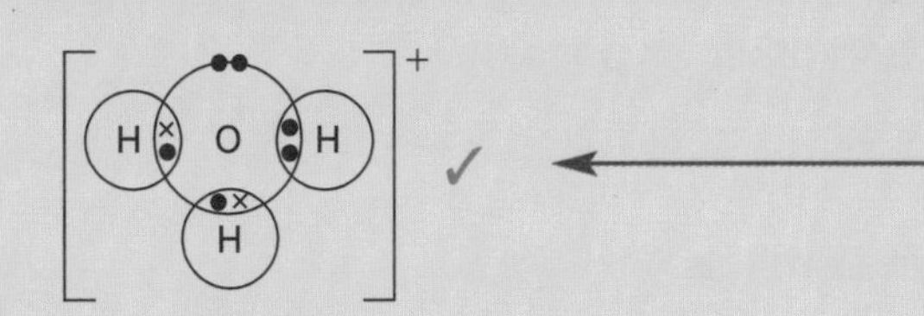

✓

Chlorine has gained an electron.

(b) Use these structures to explain the meaning of:

(i) covalent bond **(ii)** dative covalent bond. [4]

(i) A covalent bond is a shared ✓ pair of electrons. ✓

(ii) A dative covalent bond is one in which one of the atoms ✓ supplies both. ✗

You have forgotten to say 'supplies both of the shared electrons'.

(c) When ammonia (NH_3) is added to water, the following reaction occurs:

$NH_3(g) + H_2O(l) \rightleftharpoons NH_4^+(aq) + OH^-(aq)$

Use a dot and cross diagram to show the structure of the OH^- ion. [2]

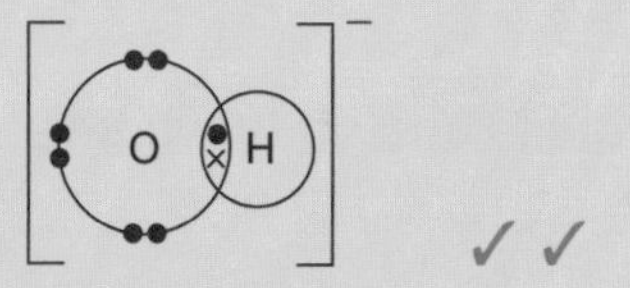

✓ ✓

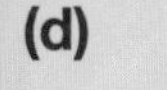

(d) **(i)** Name the ionic compound formed when hydrogen chloride reacts with ammonia. [1]

(i) ammonium chloride ✓

(ii) Draw a diagram to show the arrangement of electrons in the compound you have named in **(d) (i)**. [3]

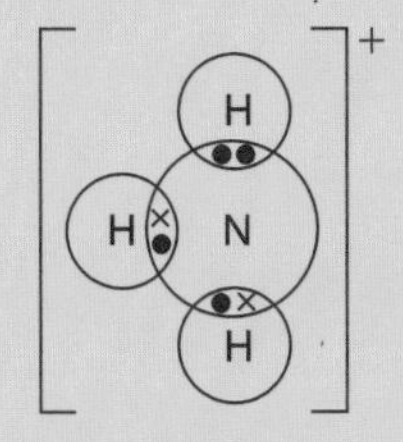

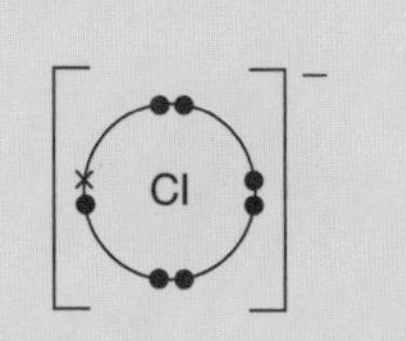

✓ ✗ ✓

You have forgotten to show the fourth hydrogen atom in NH_4^+.

(e) What are the shapes of the following: [4]

(i) hydrogen chloride **(ii)** water **(iii)** ammonia **(iv)** H_3O^+?

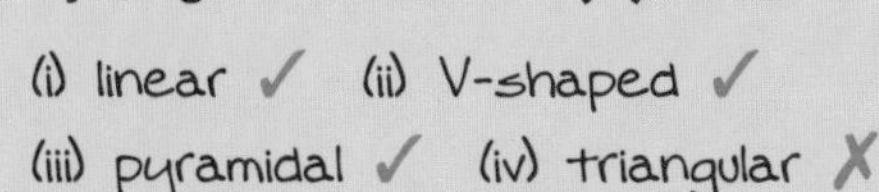

(i) linear ✓ (ii) V-shaped ✓

(iii) pyramidal ✓ (iv) triangular ✗

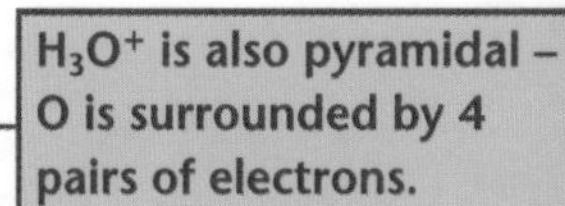

H_3O^+ is also pyramidal – O is surrounded by 4 pairs of electrons.

Exam practice questions

1 The theory of shapes of molecules was developed by Nyholm and Gillespie. They stated that 'electron pairs, whether in bonding orbitals or lone-pair orbitals, arrange themselves to be as far apart from each other as possible'.

(a) What do you understand by:

(i) bonding orbitals [1]

(ii) lone-pair orbitals? [1]

(b) Look at the following molecules:

```
      H              H              F
      |              |          F   |   F
  H—N—H        H—C—H            \ S /          H—S—H
                     |          F / | \ F
                     H              F

      A              B              C              D
```

(i) Which molecule has the greatest number of bonded pairs of electrons? [1]

(ii) Which molecule has the greatest number of lone pairs of electrons around the central atom? [1]

(iii) Identify the shape of each of the molecules and state whether it is planar or non-planar. [4]

(c) The ion NH_2^- is found in compounds such as sodium amide, $NaNH_2$.

(i) Draw a diagram showing the outer electronic structure of the nitrogen atom. [1]

(ii) Draw a 'dot-and-cross' diagram to show the arrangement of the outermost electrons in NH_2^-. [1]

(iii) State, with reasons, the shape of the NH_2^- ion. Your answer should give an approximate value for the H–N–H angle. [3]

[Total: 13]

Answers on pages 30–31 Answers on pages 30–31 Answers on pages 30–31

2 The table below gives some values for electronegativity on the Pauling scale.

C 2.5	N 3.0	O 3.5	F 4.0
Si 1.8	P 2.1	S 2.5	Cl 3.0

The value for hydrogen is 2.1.

(a) What is meant by electronegativity? [2]

(b) The hydrogen fluoride molecule can be written as $H^{\delta+}$—$F^{\delta-}$. It is a polar molecule and has a dipole moment.

(i) What do the symbols $\delta+$ and $\delta-$ represent? [1]

(ii) What is meant by a polar bond? [1]

(iii) What is meant by 'a polar molecule'? [1]

(c) Using only the elements in the above table of electronegatives and hydrogen, complete the following table. (The first one has been completed for you.)

Description	Example	Shape	Polar bonds	Has dipole moment
pentatomic molecule	CCl_4	tetrahedral	yes	no
diatomic molecule	F_2			no
tetratomic molecule		pyramidal		yes
	CO_2	linear	yes	
triatomic molecule	H_2O		yes	
		octahedral		no

[5]

[Total: 10]

Structure and bonding

Answers on pages 30–31 Answers on pages 30–31 Answers on pages 30–31

3 **(a)** Outline how van der Waals' forces arise in covalent compounds. [3]

(b) The relative strengths of the van der Waals' forces for various molecules are given in the table below.

Molecule	**Number of electrons in molecule**	**Relative van der Waals' force**	**Boiling point °C**
hydrogen H_2	2	11.3	−253
carbon monoxide CO	14	67.0	−191
hydrogen chloride HCl	18	105.0	−85
chlorine Cl_2		461.0	−35

(i) How many electrons are there in a chlorine molecule? [1]

(ii) Make two deductions from the above table. [2]

(iii) The corresponding figures for water are:

Molecule	**Number of electrons in molecule**	**Relative van der Waals' force**	**Boiling point °C**
water H_2O	10	47	100

Suggest why water does **not** fit the above pattern. [2]

(c) BCl_3 and NCl_3 are covalent compounds that both have simple molecular structures.

(i) Draw dot and cross diagrams to show the electronic structures of a molecule of each compound. (You need only show the outer electrons of the atoms.) [2]

(ii) Explain why a molecule of BCl_3 is non-polar whereas a molecule of NCl_3 is polar. [2]

(iii) NCl_3 has a boiling point of 71 °C and BCl_3 has a boiling point of 15 °C. Suggest reasons for the difference in boiling points. Explain your reasoning. [3]

[Total: 15]

Answers on pages 30–31 Answers on pages 30–31 Answers on pages 30–31

4 Hydrogen can form a covalent bond as in hydrogen chloride, an ionic bond as in sodium hydride, a dative covalent bond as in the ammonium ion and hydrogen bonding as in water.

(a) Using the above compounds as examples explain what is meant by:

(i) a covalent bond [2]

(ii) an ionic bond [2]

(iii) a dative covalent bond [2]

(iv) hydrogen bonding. [2]

(b) Which of the following molecules have hydrogen bonding?

H_2S CO_2 SO_2 CH_4 NH_3 HF CH_3OH [3]

(c) Give **two** properties of water arising from hydrogen bonding. [2]

(d) Ethanoic acid dimerises to form the 'molecule' $(CH_3COOH)_2$. The molecule contains hydrogen bonding. The diagram below shows a molecule of ethanoic acid.

By adding another molecule of ethanoic acid, suggest the hydrogen bonding that exists between the two molecules.

```
        O
       //
H3C – C
       \
        O – H
```

[2]

[Total: 15]

Structure and bonding

Answers on pages 30–31 Answers on pages 30–31 Answers on pages 30–31

Answers

(1) (a) (i) A pair of electrons, shared between two atoms, with one electron from each orbital. ✓

(ii) A pair of electrons in the same orbital, not involved in bonding. ✓

(b) (i) C: sulphur hexafluoride ✓

(ii) D: hydrogen sulphide ✓

(iii) A: pyramidal, non-planar ✓ B: tetrahedral, non-planar ✓
C: octahedral, non-planar ✓ D: V-shaped (non-linear), planar ✓

examiner's tip

When defining shapes, the lone pairs are ignored. Read the question carefully. Sometimes you are asked how the pairs of electrons are arranged: the answer for both ammonia and hydrogen sulphide would be tetrahedral.

(c) (i) •N• (with five electrons shown as dots) ✓

(ii) $[H \; N \; H]^-$ (dot-and-cross diagram) ✓
electron from metal

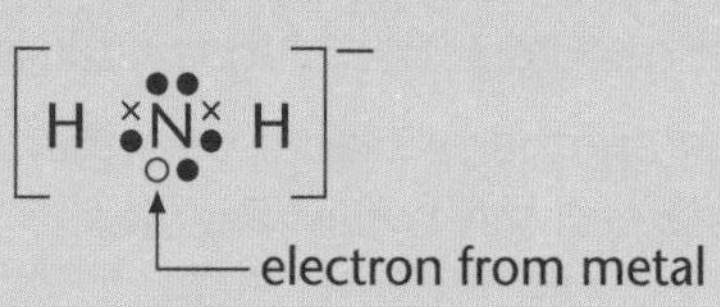

(iii) V-shaped (non-linear), ✓ four pairs of electrons would arrange themselves tetrahedrally. Two are lone pairs and therefore non-linear. ✓ The angle will be between 104° and 106°. ✓

examiner's tip

The shape of NH_2^- will be very similar to that of water. Non-bonded pairs of electrons repel one another more than bonded pairs. Hence the angle is less than the tetrahedral angle of 109.5° or 107° from one non-bonded pair of electrons.

---

(2) (a) It is a measure of the attraction of an atom in a molecule ✓ for the pair of electrons in a covalent bond. ✓

(b) (i) small positive charge and a small negative charge ✓

(ii) the electrons in the bond are shared unequally ✓

(iii) a molecule that will move when placed in a magnetic or electric field ✓

examiner's tip

For a molecule to be polar, the bonds must be polar and the molecule must not be symmetrical.

(c)

Description	Example	Shape	Polar bonds	Has dipole moment	
pentatomic molecule	CCl_4	tetrahedral	yes	no	
diatomic molecule	F_2	*linear*	*no*	no	✓
tetratomic molecule	*NH_3*	pyramidal	*yes*	yes	✓
triatomic molecule	CO_2	linear	yes	*no*	✓
triatomic molecule	H_2O	*non-linear*	yes	*yes*	✓
heptatomic (7 atoms) molecule	*SF_6*	octahedral	*yes*	no	✓

(3) (a) Movement of electrons produces an oscillating dipole. ✓ The oscillating dipole induces a dipole in a neighbouring molecule. ✓ The induced dipoles attract one another. ✓

> **examiner's tip**
>
> **Make sure you understand what your specification says about van der Waals' forces. Most specifications distinguish between 'dipole–dipole' interactions as if they are distinct from van der Waals' forces (dispersion forces). A hydrogen bond is a special type of strong dipole–dipole interaction.**

(b) (i) 34 ✓

(ii) As the number of electrons increases, the relative van der Waals' forces increase. ✓ As the number of electrons increases, the boiling points increase. ✓

(iii) Water has hydrogen bonding ✓ which is a much stronger bond than the van der Waals' forces. ✓

> **examiner's tip**
>
> **As a general rule, the greater the number of electrons in a molecule the higher the boiling point.**

(c) (i)

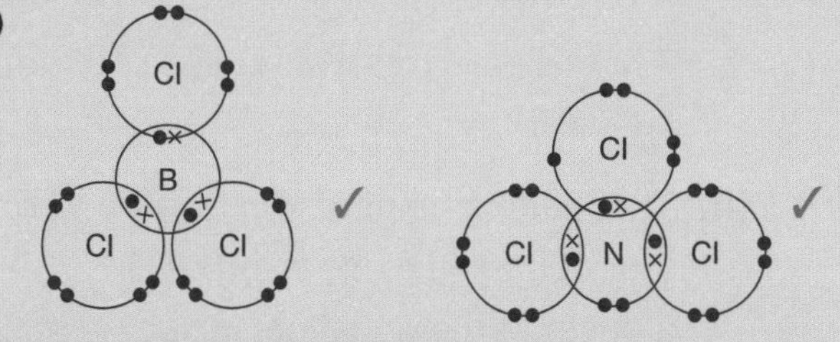

(ii) BCl_3 is symmetrical. ✓
The dipoles cancel each other out – the molecule is non-polar. ✓

(iii) The polar bond in NCl_3 is stronger since nitrogen is more electronegative than boron. ✓ Also, BCl_3 is symmetrical and dipoles will cancel. ✓ Thus the dipole–dipole interaction between the NCl_3 molecules will be greater and more energy will be required to separate the molecules. ✓

---

(4) (a) (i) A shared pair of electrons ✓ in HCl. ✓

(ii) Electrical attraction between oppositely charged ions ✓ in Na^+ H^-. ✓

(iii) A shared pair of electrons comes from just one of the bonded atoms (N:) ✓ in NH_4^+. ✓

(iv) Dipole–dipole interaction in which hydrogen is bonded to a highly electronegative element such as O, N or F ✓ in H_2O. ✓

> **examiner's tip**
>
> **Make sure that you are familiar with the four types of bonding.**

(b) NH_3, ✓ HF ✓ and CH_3OH ✓

> **examiner's tip**
>
> **All compounds with an –OH group show hydrogen bonding.**

(c) higher boiling point than expected; ✓ ice is less dense than liquid water ✓ (Other properties include, high surface tension, relatively high viscosity, capillary attraction).

(d)

```
      O----H—O
     //        \
H3C –C          C–CH3   ✓✓
     \         //
      O–H-----O
```

Chapter 4

The Periodic Table

Questions with model answers

C grade candidate – mark scored 6/10

Examiner's Commentary

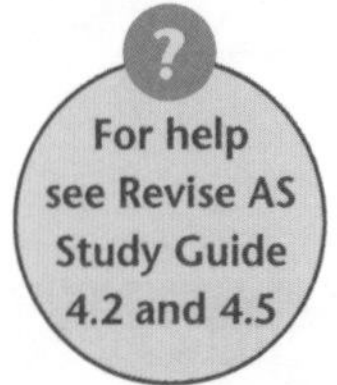

Iron is made by the reduction of iron ore, haematite, with coke. Haematite contains iron(III) oxide (Fe_2O_3).

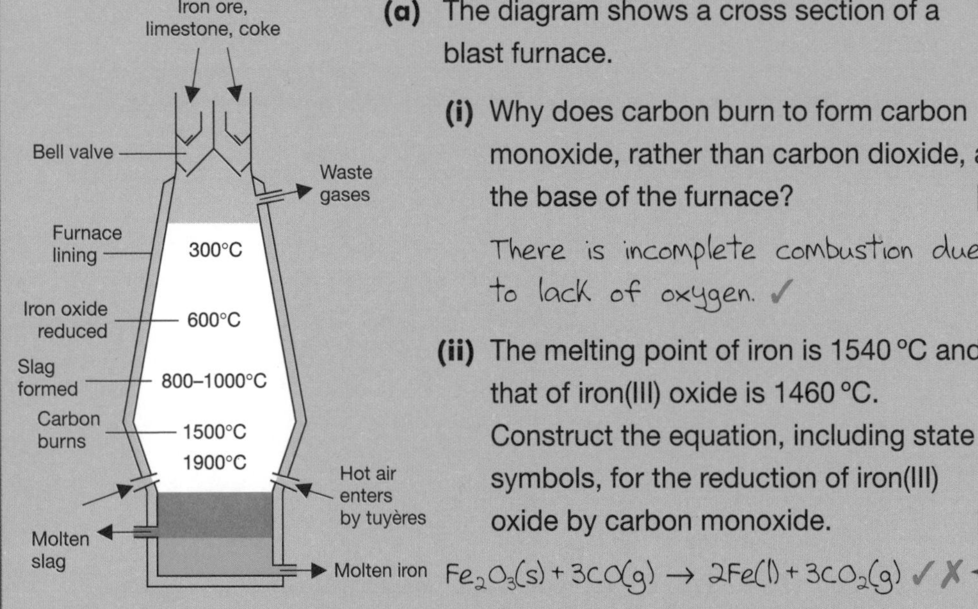

(a) The diagram shows a cross section of a blast furnace.

(i) Why does carbon burn to form carbon monoxide, rather than carbon dioxide, at the base of the furnace? **[1]**

There is incomplete combustion due to lack of oxygen. ✓

(ii) The melting point of iron is 1540 °C and that of iron(III) oxide is 1460 °C. Construct the equation, including state symbols, for the reduction of iron(III) oxide by carbon monoxide. **[2]**

$Fe_2O_3(s) + 3CO(g) \rightarrow 2Fe(l) + 3CO_2(g)$ ✓ ✗

> The temperature for the reduction is 800 °C. At this temperature, iron is a solid.

(iii) Why do the waste gases contain sulphur dioxide and nitrogen? **[2]**

The iron contains sulphur impurities ✓ *and nitrogen is from the air.* ✓

(iv) Suggest **one** advantage of slag floating on top of the iron. **[1]**

It keeps the iron molten. ✗

> It prevents the hot air reacting with the molten iron.

(b) The air entering the base of the furnace is heated by the waste gases. Give **two** advantages of heating the air in this way. **[2]**

Hot air speeds up the reaction ✓ *and it is more economical because energy is saved.* ✓

(c) Another reaction in which iron is formed is **[2]**

$$2Fe_2O_3(l) + 3C(s) \rightarrow 4Fe(l) + 3CO_2(g)$$

In terms of oxidation number, explain why this is a redox reaction.

Iron is reduced from +6 to 0 ✗ *and carbon is oxidised from 0 to +2.* ✗

> Iron is reduced from +3 to 0 and carbon is oxidised from 0 to +4. Oxidation numbers are stated for one atom only. Oxygen is always −2 (except in F_2O); therefore all other elements, when they combine with oxygen, have a positive oxidation number.

GRADE BOOSTER

Learn the oxidation rules on page 69 of the *Revise AS Study Guide*. Make sure you can apply these rules.

A grade candidate – mark scored 8/10

Examiner's Commentary

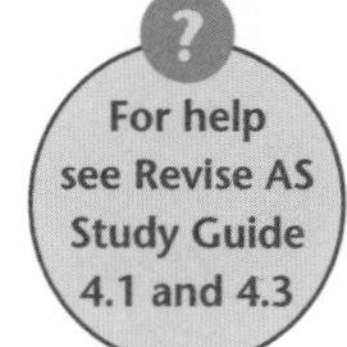

The table below gives some information about elements in Group 1 and Group 2.

	Group 1			Group 2		
	Li	Na	K	Be	Mg	Ca
Electronegativity	1.0	0.9	0.8	1.5	1.2	1.0
Ionic radius/nm	0.068	0.098	0.133	0.030	0.065	0.094
Boiling point/°C	1330	890	774	2477	1110	1487

(a) Why are Groups 1 and 2 known as the 's block' elements? **[2]**

Their electronic structures end in either s^1 ✓ or s^2. ✓

Groups 3, 4, 5, 6, 7 and 0 are the p block elements.

(b) Why are the metals in Group 1 known as the alkali metals? **[1]**

They react with water to form alkalis. ✓

Group 2 are known as the alkaline earth metals.

(c) Why does the atomic radius decrease across a period but increase down a group? **[3]**

Across a period the nuclear charge increases, thus increasing the attraction between the nucleus and the outer electrons. ✓ Extra shells are added down a group and these are further from the nucleus. ✓ The more shells the greater the shielding ✓ (both contribute to a decrease in the attraction between the nucleus and the electrons).

(d) Suggest why the boiling point:

(i) decreases going down Group 1 **[2]**

The strength of a metallic bond decreases down a group - the ions get larger and hence the attraction of the 'sea of electrons' to the nucleus decreases. ✓

(ii) increases across a period.

More electrons contribute to the 'sea of electrons', the charge on the metal ion is greater and the ion is smaller ✓ (thus increasing the attraction).

(e) The action of heat on lithium compounds is similar to the action of heat on magnesium compounds. The reason is that both lithium ions and magnesium ions have high charge densities. **[2]**

(i) What is meant by charge density?

It is the charge on the ion. ✗

It is the ratio of charge on ion : atomic radius. The larger this value, the more it will cause large anions to decompose.

(ii) Lithium carbonate decomposes to give lithium oxide and carbon dioxide. Write the equation for the action of heat on lithium carbonate.

$LiCO_3(s) \rightarrow LiO(s) + CO_2(g)$. ✗

Lithium is in Group 1. $Li_2CO_3(s) \rightarrow Li_2O(s) + CO_2(g)$

The Periodic Table

Exam practice questions

1 Fluorine is the first member of Group 7, the halogens. It is the most reactive halogen. When it reacts with elements it can bring out their highest oxidation number (which is also the group number of the element).

(a) Name **one** element that does **not** react with fluorine. [1]

(b) Write down the formula of the compound formed between the following, in which the first named element has its highest oxidation number. [2]

(i) boron and fluorine

(ii) sulphur and fluorine

(c) What is the oxidation state of oxygen in F_2O? Explain your answer. [2]

(d) Fluorine reacts with water to give hydrofluoric acid and oxygen. Write the equation for this reaction and state, with reasons, what has been oxidised and what has been reduced. [3]

(e) Give **two** reasons why fluorine has a lower boiling point than chlorine. [2]

(f) Suggest why hydrogen fluoride has a much higher melting point and boiling point than the other hydrogen halides. [2]

(g) Suggest how fluorine could be manufactured from sodium fluoride. [1]

[Total: 13]

Answers on pages 38–39 Answers on pages 38–39 Answers on pages 38–39

2 The table below gives some information about certain elements in period 3 of the Periodic Table.

	Sodium	X	Aluminium	Silicon	Y	Sulphur	Chlorine
Atomic number	11	12	13	14	15	16	17
Relative atomic mass	23	24	27	28	31	32	35.5
Boiling point /°C	890	1110	2470	2360	473	445	−34.7
Electrical conductivity	good	good	good	slight	poor	poor	poor
Formula of chloride	NaCl		$AlCl_3$	$SiCl_4$		S_2Cl_2	Cl_2

(a) For elements **X** and **Y**:

(i) write their symbols [1]

(ii) suggest the formulae of their chlorides. [2]

(b) What is the electronic structure of element 14 in terms of s and p orbitals? [1]

(c) What type of structure and physical properties are present in:

(i) sodium, **X** and aluminium [2]

(ii) silicon [2]

(iii) **Y**, sulphur and chlorine? [2]

Give reasons for your answer.

(d) The relative molecular mass of **Y** is 124. What is the molecular formula of **Y**? [1]

(e) Sodium reacts with hydrogen to form a white solid, sodium hydride NaH (Na^+H^-), melting point 800 °C.

(i) What type of bonding is present in sodium hydride? [1]

(ii) What is the oxidation number of hydrogen in sodium hydride? [1]

(iii) If molten sodium hydride is electrolysed what product would be formed at the anode? [1]

[Total: 14]

Answers on pages 38–39 Answers on pages 38–39 Answers on pages 38–39

3 A Group 2 element; calcium reacts rapidly with hydrochloric acid as shown in reaction **1** below.

$Ca(s) + 2HCl(aq) \rightarrow CaCl_2(aq) + H_2(g)$ reaction **1**

(a) Use oxidation numbers

(i) to show whether Ca, H or Cl has been oxidised in this reaction [2]

(ii) identify, with a reason, the oxidising agent in the above reaction. [2]

(b) An ionic equation can be written to show the role of $H^+(aq)$ in reaction **1**.

(i) What does $H^+(aq)$ represent? [1]

(ii) Write an ionic equation for reaction **1**. [1]

(c) Calcium reacts with water forming a solution **X** and a gas.

(i) Write an equation for the reaction of calcium with water. [2]

(ii) Predict, with a reason, the pH of the resulting solution. [1]

(iii) If solution **X** is left exposed to the air for a few days, it goes cloudy.

Explain this observation with the aid of an equation. [2]

(d) Metal **Y** is also in Group 2 of the Periodic Table and has a greater relative atomic mass than calcium. Metal **Y** also reacts with hydrochloric acid and with water.

Predict whether **Y** would react quicker, slower or at the same rate as calcium. Explain your answer. [3]

[Total: 14]

Answers on pages 38–39 Answers on pages 38–39 Answers on pages 38–39

4 The diagram below shows how the atomic radii (smooth line) and ionic radii (dotted line) vary in the Periodic Table.

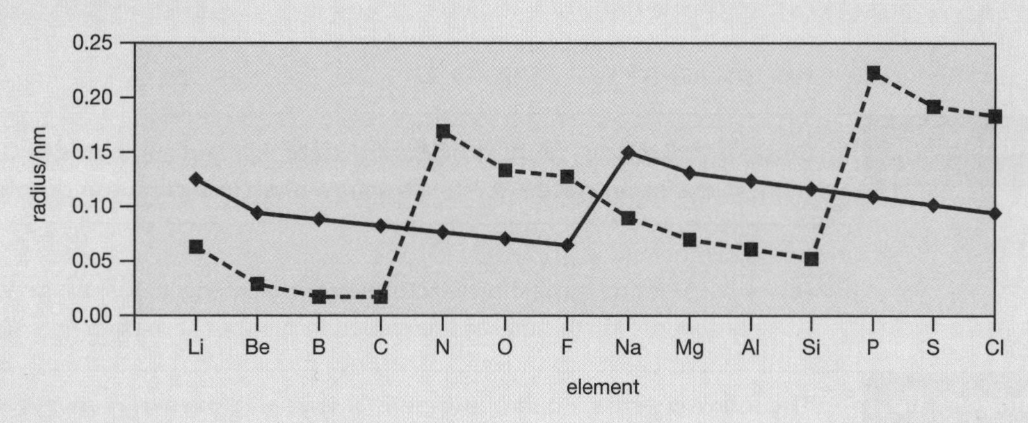

(a) The above graphs are said to be periodic.

(i) What is meant by 'periodic'? [2]

(ii) Give **three** other properties of elements that are periodic. [3]

(b) Explain the shape of the:

(i) atomic radii graph [6]

(ii) ionic radii graph. [4]

(c) From the graphs, predict the value for the radius of:

(i) potassium (K) [1]

(ii) the calcium ion Ca^{2+}. [1]

(d) Suggest why there is also a value of 0.026 nm for an ion of chlorine. [2]

[Total: 19]

The Periodic Table

Answers on pages 38–39 Answers on pages 38–39 Answers on pages 38–39

Answers

(1) (a) any noble gas ✓

(b) (i) BF_3 ✓ **(ii)** SF_6 ✓

examiner's tip

Boron is in Group 3, highest oxidation state +3, and sulphur is in Group 6, highest oxidation state +6. The structure of SF_6 is a common question.

(c) +2 ✓ Fluorine is the most electronegative element, so it always has an oxidation number of −1; this makes the oxidation number of oxygen +2. ✓

examiner's tip

The following rules apply: fluorine is always −1; oxygen is always −2, except in F_2O and peroxides (in Na_2O_2 it is −1); hydrogen is always +1, except in metal hydrides when it is −1. Note the spelling of fluorine.

(d) $2F_2(g) + 2H_2O(l) \rightarrow 4HF(aq) + O_2(g)$ ✓
Fluorine has been reduced, oxidation number reduced from 0 to −1. ✓
Oxygen has been oxidised, oxidation number increased from −2 to 0. ✓

(e) Fluorine has fewer electrons ✓ and weaker van der Waals forces. ✓

(f) Hydrogen bonding. ✓ Hydrogen fluoride is a polar molecule with a lone pair of electrons on F, which is highly electronegative. ✓

examiner's tip

Other compounds that contain hydrogen bonding are water and ammonia.

(g) by electrolysis ✓

---

(2) (a) (i) Mg and P ✓ **(ii)** $MgCl_2$ ✓ and either PCl_3 or PCl_5 ✓

(b) $1s^2\ 2s^2\ 2p^6\ 3s^2\ 3p^2$ ✓

examiner's tip

Silicon is a p block element in Group 4 of the Periodic Table.

(c) (i) Giant metallic. ✓ They have a high boiling point and conduct ✓ when solid.

(ii) Giant atomic. ✓ It has a high boiling point, and is a semi-conductor. ✓

(iii) Simple molecular. ✓ They have a low boiling point and do not conduct. ✓

(d) $\frac{124}{31} = 4$ $\mathbf{Y}_4$ ✓

(e) (i) ionic ✓ **(ii)** −1 ✓ **(iii)** hydrogen ✓

examiner's tip

Note, it is molten sodium hydride that is electrolysed. This is the only time hydrogen appears at the anode – normally it is released (from H^+ ions) at the cathode.

(3) (a) (i) Ca is oxidised, ✓ increasing in oxidation number from 0 to +2 ✓

(ii) H^+ ion; ✓ its oxidation number decreases from +1 to 0 ✓

examiner's tip

This question asked you to use oxidation number. You would have received little credit for answering the question in terms of electron gain and loss.

(b) (i) hydrated hydrogen ions (an acid) ✓

(ii) $Ca(s) + 2H^+(aq) \rightarrow Ca^{2+}(aq) + H_2(g)$ ✓

(c) (i) $Ca(s) + 2H_2O(l) \rightarrow Ca(OH)_2(aq) + H_2(g)$ ✓✓ (for state symbols)

(ii) pH approximately 11; it is a strong alkali ✓

(iii) CO_2 from the air reacts with calcium hydroxide solution ✓ to form calcium carbonate which is insoluble.

$$Ca(OH)_2(aq) + CO_2(g) \rightarrow CaCO_3(s) + H_2O(l)$$ ✓

(d) Y would react more quickly than calcium. ✓ Elements in Groups 1 and 2 get more reactive going down the group because the atoms get larger ✓ and they lose their outermost electrons more readily. ✓

---

(4) (a) (i) A property of the elements that repeats periodically ✓ when the elements are arranged in order of their atomic numbers. ✓

(ii) Melting point and boiling points, ✓ first ionisation energies ✓ and electrical conductivities. ✓ (Other possibilities include electronegativity and atomic volume.)

(b) (i) Across a period, nuclear charge increases ✓ as outer electrons are being added to the same shell; ✓ the attraction between the nucleus and the electrons increases making the atomic radius decrease in value. ✓

Going down the group, extra shells are added that are further from the nucleus; ✓ there is increased shielding of the outer electrons from the nucleus. ✓

Both factors make the attraction between the outer electrons and the nucleus less and therefore atomic radius increases. ✓

(ii) The explanation for the ionic radius is similar. When an element loses electrons, it loses a shell and there is a stronger attraction between the electrons and the nucleus. ✓ This makes the ionic radius of an element smaller than the atomic radius.

The more electrons lost in forming positively charged ions, the smaller the ion, e.g. $Li^+ > Be^{2+} > B^{3+} > C^{4+}$. ✓

When atoms gain electrons, the extra electron in the outer energy level is repelled making the radius of the ion larger. ✓

The more electrons added the larger the ionic radius, e.g. $N^{3-} > O^{2-} > F^-$. ✓

(c) (i) 0.20 nm to 0.25 nm ✓ **(ii)** 0.15 nm to 0.19 nm ✓

examiner's tip

(i) should be larger than the atomic radius of Na and (ii) should be larger than the ionic radius of Mg^{2+}. This is because there is an extra shell in each case.

(d) Since this value (0.026 nm) is smaller than the value for the atomic radius, ✓ then chlorine must be able to form a positively charged ion. ✓

Chapter 5 **Chemical energetics**

Questions with model answers

C grade candidate – mark scored 7/12

Examiner's Commentary

(a) State Hess's Law. [3]

If a reaction can take place by more than one route ✓ and the initial and final conditions are the same, ✓ the total enthalpy change is the same for each route. ✓

? For help see Revise AS Study Guide 5.2

(b) In the diagram below, give the formulae of **X** and **Y**. [2]

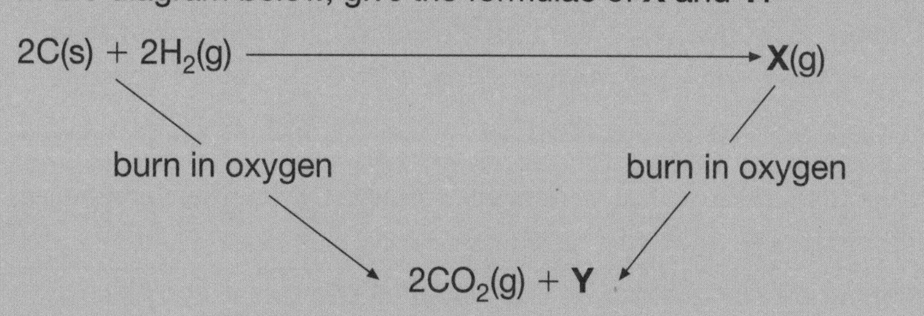

X is C_2H_4 ✓ **Y** is H_2 hydrogen ✗

Hydrogen, in the presence of oxygen, always forms water.

(c) The standard enthalpy change of combustion of carbon is $-394\ kJ\ mol^{-1}$, and that of carbon monoxide is $-111\ kJ\ mol^{-1}$.

Calculate the standard enthalpy change of reaction for:

$2C(s) + O_2 \rightarrow 2CO(g)$ [3]

$C(s) + \frac{1}{2}O_2(g) \xrightarrow{\Delta H_r} CO$

$+\frac{1}{2}O_2(g)$ -394 -111 $+\frac{1}{2}O_2(g)$

$CO_2(g)$ ✓

$\Delta H_r + (-111) = -394$ ✓ $\Delta H_r = -283\ KJ\ mol^{-1}$ ✗

You have forgotten that the reaction was for the production of 2 moles of CO. Answer = $-566\ kJ\ mol^{-1}$.

(d) The enthalpy change of combustion of magnesium is $-602\ kJ\ mol^{-1}$. Would you expect magnesium to burn in carbon dioxide to form magnesium oxide and carbon? Explain your answer. [4]

The equation would have been:

$2Mg(s) + CO_2(g) \rightarrow 2MgO(s) + C(s)$ ✓

but carbon dioxide does not support burning ✗✗✗

You should have used the enthalpy change of combustion of carbon to show that the reaction is $-810\ kJ\ mol^{-1}$ exothermic, and is therefore likely to take place.

GRADE BOOSTER

There were 4 marks for part (d) and the question gave the enthalpy change of combustion of magnesium. This implies that more was required than the answer given by the student.

A grade candidate – mark scored 8/10

Examiner's Commentary

(a) The average healthy person needs 12 500 kJ from sugars such as sucrose ($C_{12}H_{22}O_{11}$) a day. The enthalpy change of combustion of sucrose is $-5670\ kJ\ mol^{-1}$. How many kilograms of sucrose (to three decimal places) must be eaten each day to provide this energy? **[3]**

M_r of sugar is 342 (12 × 12 + 22 × 1 + 11 × 16) ✓
5670 kJ is produced by 342 g ✓

12 500 kJ is produced by $\frac{342 \times 12\,500}{5670}$ = 754 g = 0.754 kg ✓

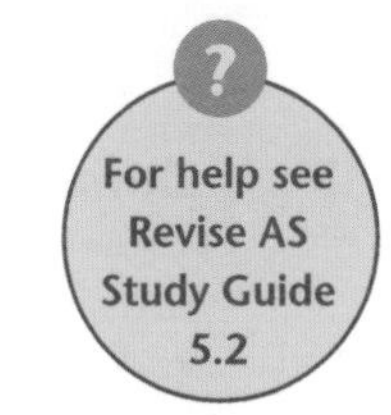

Remember many foods contain sugar.

(b) It is too dangerous to measure the enthalpy change for the reaction:

$ZnO(s) + Mg(s) \rightarrow Zn(s) + MgO(s)$

It is also difficult to measure the enthalpy changes of combustion of zinc and magnesium. The enthalpy change can be worked out indirectly by a series of experiments. One of the experiments would be to work out the enthalpy change for the reaction:

$MgO(s) + 2HCl(aq) \rightarrow MgCl_2(aq) + H_2O(l)$ reaction 1

(i) Write the equations for the other experiments you would carry out. Show how you would use the results to calculate the enthalpy change for the reaction between zinc oxide and magnesium. **[3]**

$ZnO(s) + 2HCl(aq) \rightarrow ZnCl_2(aq) + H_2O(l)$ ✓ reaction 2
$Mg(s) + ZnCl_2(aq) \rightarrow MgCl_2(aq) + Zn(s)$ ✓ reaction 3
add together the enthalpy changes for 2 and 3 and subtract those for 1. ✓

This is an application of Hess's Law where an energy change can be calculated indirectly.

(ii) Why cannot the enthalpy change for

$CuO(s) + Mg(s) \rightarrow Cu(s) + MgO(s)$

be calculated by a similar method? **[1]**

Because CuO does not react with hydrochloric acid. ✗

Copper does NOT react with hydrochloric acid.

(c) The reaction below is used to weld together sections of railway lines.

$2Al(s) + Fe_2O_3(s) \rightarrow 2Fe(s) + Al_2O_3(s)$

The standard enthalpy changes of formation of Fe_2O_3 and Al_2O_3 are $-836\ kJ\ mol^{-1}$ and $-1664\ kJ\ mol^{-1}$ respectively.

(i) Calculate the standard enthalpy change of this reaction. **[2]**

Enthalpy change = −1664 − (−836) ✓ = −828 kJ mol⁻¹. ✓

(ii) Suggest why this reaction can be used to weld materials made of iron. **[1]**

Iron is formed. ✗

The energy produced is sufficient to melt the iron.

Exam practice questions

1

Definition of enthalpy change	Type of enthalpy change	Enthalpy change
required to break and separate 1 mole of bonds in the molecules of a gaseous element or compound so that the resulting gaseous species exert no forces upon each other	bond enthalpy	**A**
B	standard enthalpy change of neutralisation	−ve
C	standard enthalpy change of combustion	**D**
when one mole of a compound in its standard state is formed from its constituent elements in their standard states under standard conditions	**E**	usually −ve

(a) Identify **A**, **B**, **C**, **D**, and **E** in the above table. [8]

(b) The enthalpy change for the reaction $2Mg(s) + O_2(g) \rightarrow 2MgO(s)$ is $-1204\ kJ\ mol^{-1}$.
What will be the value for the enthalpy change **E** of magnesium oxide?
Explain your answer. [2]

(c) What is meant by the standard state for an element? [2]

(d) **(i)** The standard enthalpy change of neutralisation for the reaction between aqueous sodium hydroxide and hydrochloric acid is the same as for the reaction of aqueous potassium hydroxide and nitric acid.
Explain why this is so. [3]

(ii) Suggest why the value is less exothermic for the reaction between hydrofluoric acid (a weak acid) and sodium hydroxide. [1]

[Total: 16]

Answers on pages 46–47 Answers on pages 46–47 Answers on pages 46–47

2 The bond energies, in kJ mol^{-1}, below, show the successive energies required to break the bonds in methane, water and carbon dioxide.

Methane		**Water**		**Carbon dioxide**	
$H_3C–H$	+425	HO–H	+494	OC=O	+531
$H_2C–H$	+470	H–O	+430	O=C	+1075
HC–H	+416				
C–H	+335				

The bond enthalpy of O=O is +497 kJ mol^{-1}.

(a) **(i)** Calculate the mean bond enthalpies in methane, water and carbon dioxide. [2]

(ii) Hence calculate the enthalpy change of combustion of methane (CH_4) forming carbon dioxide and steam. [4]

(b) Carbon burns in an incomplete supply of oxygen to form carbon monoxide. The standard enthalpy change of atomisation ΔH_a is the enthalpy change when a substance, in its standard state, decomposes to form 1 mole of atoms in the gaseous state under standard conditions. (For diatomic elements it is equal to half the bond enthalpy.)

The standard enthalpy change of formation of carbon monoxide is −111 kJ mol^{-1}.

ΔH_a for carbon is +715 kJ mol^{-1} and ΔH_a for oxygen is +249 kJ mol^{-1}.

(i) From this information show that the bond enthalpy in carbon monoxide is +1075 kJ mol^{-1}. [2]

(ii) Suggest why the 'second bond' enthalpy for carbon dioxide and the bond enthalpy for carbon monoxide are the same. [1]

(c) The standard enthalpy changes of combustion for three alcohols are:

CH_3OH −726 kJ mol^{-1} C_2H_5OH −1366 kJ mol^{-1} C_3H_7OH −2017 kJ mol^{-1}

Using these figures:

(i) predict, with a reason, the standard enthalpy change of combustion of C_4H_9OH [2]

(ii) predict, with a reason, the standard enthalpy change of combustion of water and comment on your answer. [2]

[Total: 13]

Chemical energetics

Answers on pages 46–47 Answers on pages 46–47 Answers on pages 46–47

3 **(a)** When a student dissolved 4.0 g of anhydrous copper(II) sulphate in 50.0 cm^3 of water the temperature rise was 6.0 °C. An aqueous solution of copper(II) sulphate was formed.

The specific heat capacity of the solution, c, is 4.18 $J\ g^{-1}\ K^{-1}$.

(i) Calculate the enthalpy change. [1]

Ignore the mass of anhydrous copper(II) sulphate in your calculation.

(ii) Calculate the number of moles of anhydrous copper(II) sulphate used. [2]

(iii) Calculate the enthalpy change per mole of anhydrous copper(II) sulphate. State, with a reason, whether the reaction is exothermic or endothermic. [3]

(b) The student repeated the experiment by adding hydrated copper(II) sulphate, $CuSO_4.5H_2O$ to an excess of water to form an aqueous solution of copper(II) sulphate.

He calculated an enthalpy change for this reaction of +8.36 kJ mol^{-1} and constructed the enthalpy cycle below. The arrow heads have been missed off.

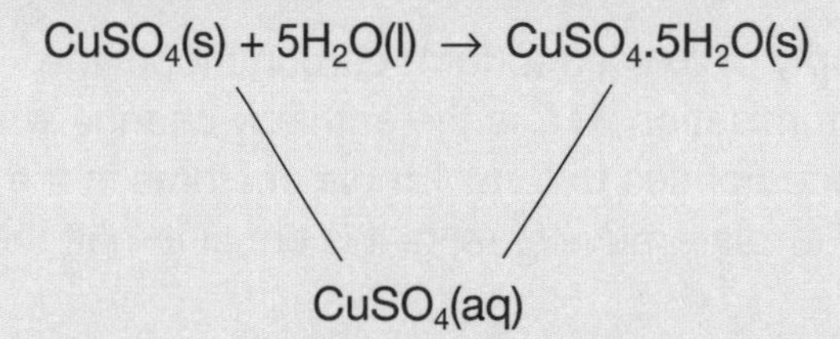

(i) Add arrow heads to complete the enthalpy cycle. [1]

(ii) Use this cycle and your answer to **(a) (iii)** to calculate the unknown enthalpy change of reaction in kJ mol^{-1}. [2]

(c) Suggest why it would be difficult to measure this heat of reaction directly. [2]

(d) Predict the temperature change if one mole of anhydrous copper(II) sulphate were added to 1 dm^3 of water. [1]

[Total: 12]

Answers on pages 46–47 Answers on pages 46–47 Answers on pages 46–47

4 **(a)** The standard enthalpy change of formation of propene (C_3H_6) is +20.4 kJ mol^{-1} and the standard enthalpy change of formation of propane (C_3H_8) is −103.8 kJ mol^{-1}.

By drawing a suitable energy cycle, calculate the enthalpy change for the reaction. State any law you use.

$C_3H_6(g) + H_2(g) \rightarrow C_3H_8(g)$ [7]

(b) The standard enthalpy of formation of propene cannot be found directly, but it can be calculated from other reactions.

State the reactions that may be used. (Details of the experiments are not required.) [1]

(c) You are supplied with the following bond enthalpies.

Bond	Average bond enthalpies /kJ mol^{-1}
C–H	+413
C=C	+612
C–C	+347
H–H	+436

Use these bond enthalpies to calculate a second value for the standard enthalpy change of the reaction $C_3H_6(g) + H_2(g) \rightarrow C_3H_8(g)$ [3]

(d) Which of the two values **(a)** or **(c)** do you think is the more accurate and why? [2]

[Total: 13]

Answers on pages 46–47 Answers on pages 46–47 Answers on pages 46–47

Answers

(1) (a) A +ve; ✓ B The enthalpy change that accompanies the neutralisation of an acid by a base ✓ to form 1 mole of $H_2O(l)$ ✓ under standard conditions.
C The enthalpy change that takes place when one mole of a substance ✓ reacts completely with oxygen ✓ under standard conditions, all reactants and products being in their standard state. ✓
D –ve; ✓ E Standard enthalpy change of formation. ✓

(b) -602 kJ mol^{-1} ✓ Enthalpy of formation is for the formation of 1 mole of MgO. ✓

(c) physical state at a pressure of 100 kPa ✓ a stated temperature (normally 298 K) ✓

(d) (i) The acids and alkalis are fully ionised. ✓
Enthalpy of neutralisation is for the formation of 1 mole of water. ✓
The reaction taking place in each case is $H^+(aq) + OH^-(aq) \rightarrow H_2O(l)$. ✓

examiner's tip

(a) Note that the definitions refer to 1 mole.

(b) Don't forget the sign + or – and units. Enthalpy change of reaction must always be quoted with an equation to show the molar amounts that are reacting. Note that this formation is exothermic.

(d) (i) Strong acids and strong alkalis ionise completely.

(ii) The value for a weak acid and a strong alkali is less because energy has to be used to ionise the weak acid. ✓

---

(2) (a) (i) methane 411.5 kJ mol^{-1}; ✓ water 462 kJ mol^{-1}; carbon dioxide 803 kJ mol^{-1} ✓

(ii) Equation is $CH_4(g) + 2O_2(g) \rightarrow CO_2(g) + 2H_2O(g)$ ✓

Energy required to break bonds = 4×411.5 (from CH_4) + 2×497 (from O_2)
= +2640 kJ ✓

Energy released by bond making = 2×803 (from CO_2) + 4×462 (from H_2O)
= −3454 kJ ✓

$\Delta H_c = -814$ kJ mol^{-1} ✓

(b) (i) $C(s) + \frac{1}{2}O_2(g) \rightarrow CO(g)$ $\Delta H_f = -111$ kJ mol^{-1}
$C(s) \rightarrow C(g)$ $\Delta H_a = +715$ kJ mol^{-1}
$\frac{1}{2}O_2(g) \rightarrow O(g)$ $\Delta H_a = +249$ kJ mol^{-1}

$CO(g) \rightarrow C(g) + O(g)$ $\Delta H = -(-111) + (715 + 249) = 1075$ kJ mol^{-1} ✓
Bond enthalpy = + 1075 kJ mol^{-1} ✓

(ii) the same bonds are being broken ✓

(c) (i) Differences in ΔH_c are 640 kJ mol^{-1} and 651 kJ mol^{-1} ✓
value for C_4H_9OH is between –2650 and –2680 kJ mol^{-1} ✓

(ii) between −80 and −90 kJ mol^{-1} ✓ expect it to be zero ✓

examiner's tip

(a) (ii) Bond breaking is endothermic; bond making is exothermic. All combustion reactions are exothermic.

(b) (ii) When the first C=O bond is broken in carbon dioxide, the structure changes to the bonding in carbon monoxide.

(c) (ii) If a CH_2 is subtracted from CH_3OH we get water. Water can be considered as an alcohol without carbon.

(3) (a) (i) Energy gain by surroundings = $50 \times 4.18 \times 6 = 1254$ Joules ✓

(ii) Moles $CuSO_4$ $\frac{4}{159.5}$ ✓ = 0.025 moles ✓

(iii) Energy **lost** by dissolving 1 mol $CuSO_4 = \frac{1254}{0.025}$ ✓ = 50 160 J mol^{-1} (50.160 kJ mol^{-1}). ✓

Enthalpy change = −50.160 kJ mol^{-1} (Note the negative sign in the **enthalpy change**)

The reaction is exothermic because the temperature rose. ✓

(b) (i) $CuSO_4(s) + 5H_2O(l) \xrightarrow{\Delta H_r} CuSO_4.5H_2O(s)$

$CuSO_4(aq)$

(two arrowheads) ✓

examiner's tip

Make sure that you know how to construct an enthalpy cycle. The arrows point in the direction of the enthalpy change you are given.

(ii) $\Delta H_r = -50.16 + (-8.36)$ ✓ = −58.72 kJ mol^{-1} ✓

(c) Some of the copper(II) sulphate would dissolve in the water added; ✓ since the reaction is exothermic some of the water would be lost as steam. ✓

(d) 12.0 °C ✓

---

(4) (a) $C_3H_6(g) + H_2(g) \xrightarrow{\Delta H_r} C_3H_8(g)$

+20.4 kJ mol^{-1} −103.8 kJ mol^{-1}

carbon and hydrogen ✓✓✓

Hess's Law states that, if a reaction can take place by more than one route and the initial and final conditions are the same, ✓ the total enthalpy change is the same for each route. ✓

route 1: +20.4 kJ mol^{-1} + ΔH_r route 2: −103.8 kJ mol^{-1}

∴ +20.4 kJ mol^{-1} + ΔH_r = −103.8 kJ mol^{-1} ✓

$\Delta H_r = -103.8 - (+20.4) = -124.2$ kJ mol^{-1} ✓

(b) enthalpy changes of combustion of carbon, hydrogen and propene ✓

(c) Bonds broken (C=C) + (H—H); bonds formed = (C—C) + 2(C—H) ✓

$\Delta H = [612 + 436] - [347 + (2 \times 413)]$ ✓ = −125 kJ mol^{-1} ✓

(d) (a) ✓ Bond energies are average values. ✓

examiner's tip

The bond energies vary according to the environment of the bond. E.g. the C=O bond energy in carbon dioxide is +803 kJ mol^{-1} but, in aldehydes such as ethanal, it is +736 kJ mol^{-1}.

Chapter 6

Rates and equilibrium

Questions with model answers

C grade candidate – mark scored 6/10

Examiner's Commentary

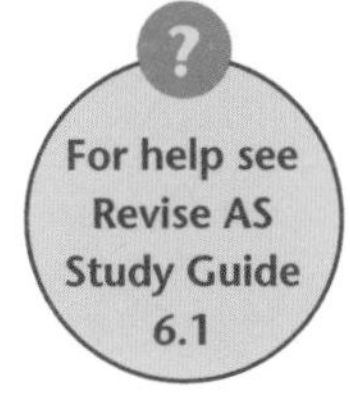

When excess lumps of barium carbonate are added to dilute hydrochloric acid the following reaction takes place:

$$BaCO_3(s) + 2HCl(aq) \rightarrow BaCl_2(aq) + H_2O(l) + CO_2(g)$$

Two experiments were carried out to follow the rate of this reaction. The results were plotted.

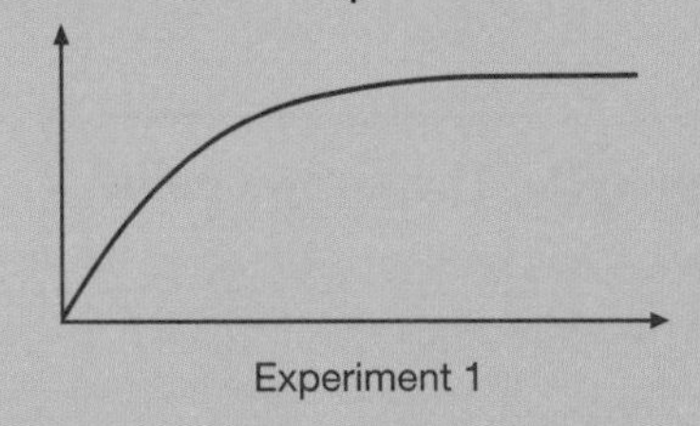
Experiment 1

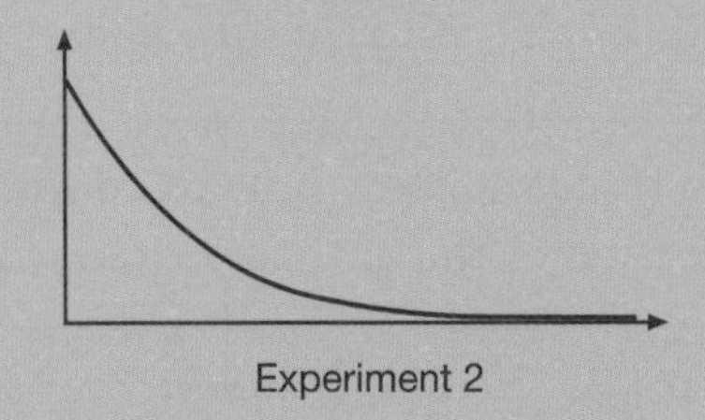
Experiment 2

(a) What would have been plotted on the *y*-axis in

(i) Experiment 1 and **(ii)** Experiment 2? [2]

(i) The volume of carbon dioxide ✓ (ii) The mass of reactants ✓

(b) State **three** ways in which the rate of this reaction could be increased. [3]

By using smaller lumps of barium carbonate, ✓ heating the acid ✓ and increasing the concentration of the acid. ✓

Catalysts speed up reactions but there is no catalyst for this reaction.

(c) If dilute sulphuric acid had been used instead of hydrochloric acid, the reaction stops almost immediately. Suggest a reason for this observation. [2]

Sulphuric acid is a weak acid. ✗✗

Barium sulphate would be formed, which is insoluble in water and would prevent any further reaction from taking place.

(d) State, with reasons, whether the total volume of carbon dioxide given off would increase, decrease or stay the same if: [2]

(i) more lumps of barium carbonate were used?

It would stay the same – barium carbonate is in excess. ✓

(ii) the experiments were performed at a higher temperature?

There would be more – temperature speeds up the rate of reaction. ✗

Although the same number of moles of CO_2 is formed, you must remember that gases expand when they are heated.

(e) The total volume of carbon dioxide given off is slightly less than the theoretical value. What can you deduce from this observation? [1]

The reaction finished early. ✗

Carbon dioxide is slightly soluble in water.

This candidate did not know the answers and guessed. It is always a good idea to write something – you are not penalised for wrong answers.

A grade candidate – mark scored 8/10

The graph below shows the Boltzmann distribution curve for the same amount of a gas sample at two different temperatures.

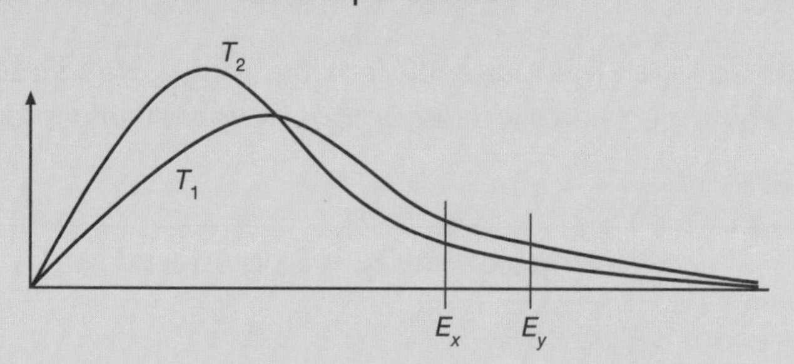

Examiner's Commentary

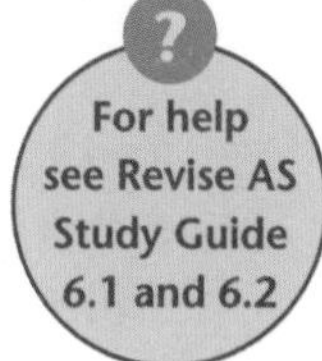

(a) **(i)** What is the label for the y-axis? [2]

Number of molecules ✓ ✗

> You should have added 'with a given energy'.

(ii) Is T_1 greater or less than T_2? Give a reason for your answer. [1]

Greater than. There are a greater number of molecules with higher energy. ✓

> The number of particles is the same, but they are distributed differently.

(iii) What do E_x and E_y represent? [2]

E_x = activation energy with a catalyst. ✓
E_y = activation energy for the reaction. ✓

(b) Explain, in terms of the collision theory, why an increase in temperature increases the rate of a chemical reaction. [2]

It brings about more collisions each second ✓ and a greater proportion of the molecules exceed the activation energy. ✓

> Notice that an increase in energy has two effects. A rise of 10 °C approximately doubles the rate of a reaction.

(c) The gas, nitrogen monoxide, NO, was used as a catalyst in the manufacture of sulphur trioxide. [3]

$$2SO_2(g) + O_2(g) \rightleftharpoons 2SO_3(g)$$

(i) Why is nitrogen monoxide a homogeneous catalyst for this reaction?

It is in the same physical state as the reactants and products. ✓

(ii) Nitrogen monoxide provides an alternative route for the reaction. It causes the reaction to take place in two stages. The second stage is

$$2SO_2(g) + 2NO_2(g) \rightarrow 2SO_3(g) + 2NO(g)$$

Write an equation for the first stage.

$2SO_2(g) + 2NO(g) \rightarrow 2SO_3 + N_2(g)$ ✗

> The second-stage equation tells you that NO_2 has been formed and the overall equation tells you that oxygen is involved. The first stage should be $2NO(g) + O_2(g) \rightarrow 2NO_2(g)$. The first and second stages added together give the overall reaction.

(iii) How does the catalyst, nitrogen monoxide, speed up the rate of the reaction?

It lowers the activation energy. ✓

Exam practice questions

1 The following results were obtained when manganese(IV) oxide, MnO_2, was added to 50 cm^3 of hydrogen peroxide solution. Hydrogen peroxide decomposes very slowly into oxygen and water at room temperature.

	lumps of MnO_2	**powdered MnO_2**
Volume of hydrogen peroxide used	50 cm^3	50 cm^3
Temperature of hydrogen peroxide	25 °C	25 °C
Mass at start	1.50 g	1.00 g
Mass at finish	1.50 g	1.00 g
Appearance at finish	powder	powder
Time to give off 25 cm^3 of oxygen	15 seconds	10 seconds

(a) Use these results to define a catalyst. [2]

(b) The decomposition of hydrogen peroxide into water and oxygen can be catalysed by adding OH^- ions. How would you show that OH^- ions were not used up in this reaction? [3]

(c) Explain why a catalyst increases the rate of a chemical reaction. [2]

(d) Write an equation to show that concentrated hydrogen peroxide is both a weak acid and a dibasic (diprotic) acid. [3]

[Total: 10]

Answers on pages 54–55 Answers on pages 54–55 Answers on pages 54–55

2 A dynamic homogeneous equilibrium is established when hydrogen and iodine react as shown.

$$H_2(g) + I_2(g) \rightleftharpoons 2HI(g)$$

(a) Use this equation to explain what is meant by:

(i) homogeneous [2]

(ii) dynamic equilibrium. [1]

(b) A mixture of the same number of moles of hydrogen and iodine was put in a container and then placed in an oven at 425 °C. The graph below shows the concentrations of hydrogen, iodine and hydrogen iodide over a period of 1 minute. (Time zero is when the gases had reached a temperature of 425 °C.)

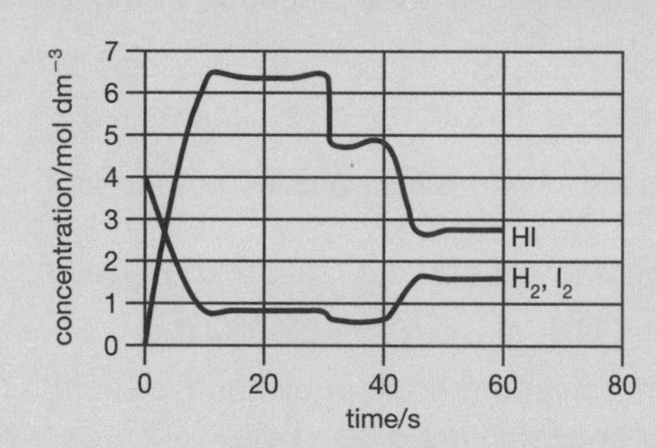

(i) Why are the molar concentrations of iodine and hydrogen the same throughout this experiment? [1]

(ii) Estimate how many seconds the reaction takes to reach equilibrium for the first time. [1]

(iii) After how many seconds was some hydrogen iodide removed from the container? [1]

(iv) After 40 seconds the oven was turned up to a much higher temperature. State with a reason whether the reaction between hydrogen and iodine is exothermic or endothermic. [2]

(c) Explain why an increase in pressure has no effect on the position of this equilibrium. [1]

(d) Iodine acts as a catalyst for the reaction between benzene and chlorine. Iodine forms an intermediate compound with chlorine. Suggest a formula for this compound. [1]

[Total: 10]

Rates and equilibrium

Answers on pages 54–55 Answers on pages 54–55 Answers on pages 54–55

3 This question looks at two different reversible reactions.

(a) **(i)** What is the meaning of the $\rightleftharpoons$ sign? [1]

(ii) State *Le Chatelier's principle*. [2]

(b) In an experiment, aqueous solutions of iron(III) ions, $Fe^{3+}(aq)$, and thiocyanate ions, $CNS^{-}(aq)$ were mixed together. A reaction took place which reached the equilibrium below.

$$Fe^{3+}(aq) + CNS^{-}(aq) \rightleftharpoons FeCNS^{2+}(aq) \qquad \Delta H \text{ −ve}$$

The solutions of $Fe^{3+}(aq)$ and $CNS^{-}(aq)$ are colourless. The colour of $FeCNS^{2+}(aq)$ is blood red.

Explain, in terms of Le Chatelier's principle, each of the following observations.

(i) When a crystal of potassium thiocyanate is added to the equilibrium mixture, the colour goes deeper red as the crystal dissolves. [2]

(ii) When the equilibrium mixture is heated, the solution becomes lighter in colour. [2]

(c) Ammonia (NH_3) is manufactured by the reaction between nitrogen and hydrogen.

$$N_2(g) + 3H_2(g) \rightleftharpoons 2NH_3(g) \quad \Delta H = -92.0\,\text{kJ mol}^{-1}$$

The percentage of ammonia formed at equilibrium depends on the temperature and pressure used. This is shown in the graph below.

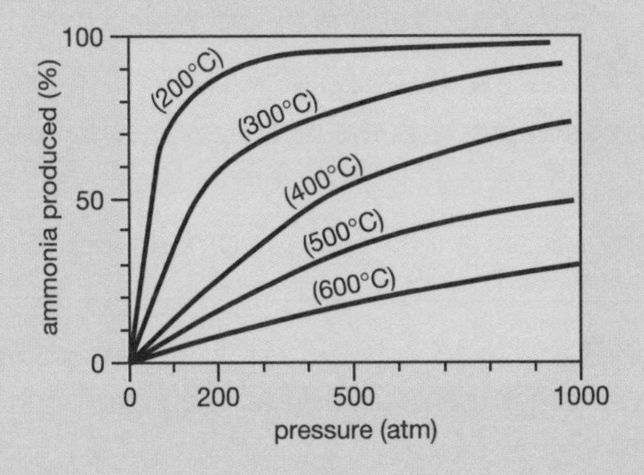

(i) Why does an increase in pressure at a constant temperature increase the % of ammonia produced? [2]

(ii) Estimate the % of ammonia produced at 200 atmosphere pressure and 200 °C. [1]

(iii) The conditions used in industry are 200 atmosphere pressure and 600 °C. Explain why these conditions are used. [3]

(iv) State one other way that the equilibrium yield of ammonia can be increased at constant pressure and constant temperature. [1]

[Total: 14]

Rates and equilibrium

Answers on pages 54–55 Answers on pages 54–55 Answers on pages 54–55

4 0.69 g of a metal carbonate M_2CO_3 was added to a flask containing a large excess of dilute hydrochloric acid and the carbon dioxide given off was collected. The reaction was carried out at constant temperature. The total volume of gas given off was recorded every 30 seconds. The results are shown in the table below.

Time/min	0.0	0.5	1.0	1.5	2.0	2.5	3.0	3.5	4.0
Total volume cm^3	10	55	80	100	115	125	130	130	130

(a) Draw a diagram of the apparatus you would use to perform this experiment. [3]

(b) The acid was saturated with carbon dioxide before starting the experiment.

(i) Suggest how the acid is saturated with carbon dioxide. [1]

(ii) Why is it necessary to saturate the acid with carbon dioxide? [1]

(c) **(i)** Plot a graph of the results. [3]

(ii) At what time was the rate of reaction 45 $cm^3\ min^{-1}$? [1]

(iii) Why does the graph remain horizontal after 3.5 minutes? [1]

(d) How would adding more of the dilute hydrochloric acid affect:

(i) the rate of the reaction [1]

(ii) the total volume of carbon dioxide given off? [1]

(e) How would adding a small amount of M_2CO_3 affect:

(i) the rate of the reaction [1]

(ii) the total volume of carbon dioxide given off? [1]

(f) The equation for the reaction is shown below.

$M_2CO_3(s) + 2HCl(aq) \rightarrow 2MCl(aq) + CO_2(g) + H_2O(l)$

(i) Calculate the number of moles of carbon dioxide given off in the initial experiment. [1]

(ii) Calculate the relative formula mass of M_2CO_3. [1]

(iii) Calculate the relative atomic mass of M and the identity of M. [1]

[Total: 17]

Answers on pages 54–55 Answers on pages 54–55 Answers on pages 54–55

Answers

(1) (a) A catalyst speeds up a chemical reaction ✓ but is not used up in the reaction. ✓

(b) By titrating measured volumes with standard acid at the start of the reaction ✓ and titrating the same volume with standard acid at the end of the reaction. ✓ The two values will be the same. ✓

examiner's tip

A catalyst is not used up in a chemical reaction. It may change in its appearance. In this example manganese(IV) oxide started as lumps but ended up as a powder.

(c) A catalyst allows a reaction to take place via a different route ✓ with a lower activation energy. ✓

examiner's tip

Since the activation energy is lowered there will be more molecules with sufficient energy to react. Thus the reaction speeds up because there are more effective collisions.

(d) $H_2O_2(aq) \rightleftharpoons 2H^+(aq) + O_2^{2-}(aq)$ ✓
(for reversible sign ✓ for correct ions ✓)

---

(2) (a) (i) All the substances are in the same physical state: ✓ in this case, gases. ✓

examiner's tip

A heterogeneous catalyst is in a different phase from the reactants. In the manufacture of ammonia the catalyst is iron, which is a solid; the reactants, nitrogen and hydrogen, are gases.

(ii) The rate of the forward reaction = the rate of the reverse reaction. ✓

examiner's tip

At equilibrium, the concentrations remain constant.

(b) (i) Equimolar amounts react ✓

examiner's tip

1 mole of H_2 and 1 mole of I_2 give 2 moles of HI.

(ii) 10 seconds ✓

(iii) 30 seconds ✓

examiner's tip

The clue was that the amounts of HI, I_2 and H_2 were reduced.

(iv) The reaction between hydrogen and iodine is exothermic ✓ since less hydrogen iodide is formed at the higher temperature. ✓

(c) There are an equal number of gaseous molecules on each side of the equation. ✓

examiner's tip

Note, you must only consider gaseous substances when looking at the effect of pressure.

(d) ICl ✓

(3) (a) (i) It is an equilibrium reaction. ✓

(ii) If a system in dynamic equilibrium is subjected to a change, ✓ processes will occur to minimise this change and to restore equilibrium. ✓

(b) (i) To relieve the effect of adding excess CNS^- ions, the equilibrium moves to the right ✓ forming more of the deep red $FeCNS^{2+}(aq)$ ions. ✓

(ii) To relieve the effect of increased temperature, the equilibrium moves to the left ✓ in the endothermic direction to form the colourless $Fe^{3+}(aq)$ and $CNS^-(aq)$ ions. ✓

(c) (i) To relieve the effect of increased pressure, the equilibrium moves to the right ✓ in the direction that produces less gas molecules. ✓

(ii) 90% ✓

(iii) The temperature is sufficiently high to allow the reaction to occur at a realistic rate, ✓ but not too high to shift the equilibrium position too far to the left, reducing NH_3; ✓ a high pressure is used, but not too high to prevent high energy and building costs. ✓

(iv) Remove the ammonia as it is formed. ✓

---

(4) (a)

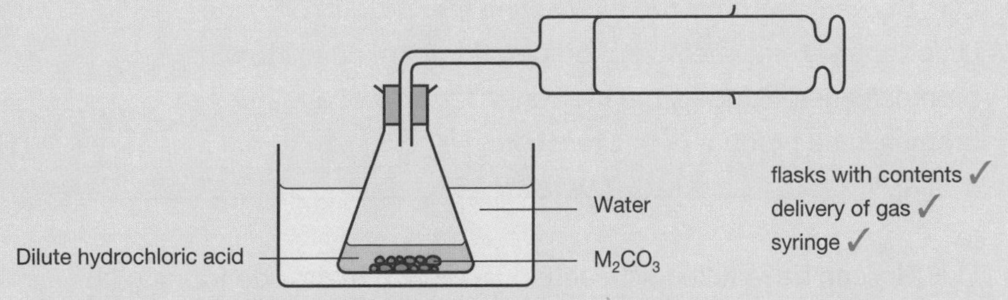

(b) (i) bubble CO_2 through acid ✓

(ii) So that none of the CO_2 given off in the experiment dissolves in the acid. ✓

(c) (i)

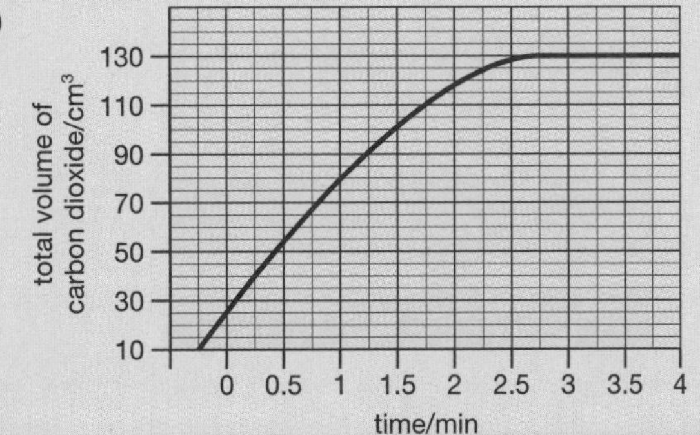

axes ✓
scale ✓
plot of curve ✓

(ii) 1 min; ✓

(iii) all the carbon dioxide has been given off ✓

(d) (i) no effect ✓ (ii) no effect ✓

(e) (i) increase rate ✓ (ii) more CO_2 given off ✓

examiner's tip

The acid is in a large excess, therefore adding more acid will have no effect, but adding more carbonate will increase the rate, greater surface area and increase the amount of CO_2 given off.

(f) (i) $\frac{120}{24\,000} = 0.005$ ✓ (ii) 138 ✓ (iii) 39, potassium K ✓

Chapter 7 Organic chemistry

Questions with model answers

C grade candidate – mark scored 9/14

(a) The compound $C_2H_4Br_2(l)$ can be made by reacting ethene with bromine.

(i) Write the equation, together with state symbols, for this reaction. [2]

$C_2H_4(g) + Br_2(l) \rightarrow C_2H_4Br_2(l)$ ✓ ✓

(ii) What are the conditions for this reaction to take place? [1]

Room temperature ✓

(iii) What colour change would you see in this reaction? [2]

The bromine is decolourised. ✓ ✗

(b) $C_2H_4Br_2$ can also be made by reacting ethane with bromine. Unfortunately, this method of preparation forms a mixture of bromoalkanes. Suggest the molecular formula of another bromoalkane product of this reaction. [1]

$C_2H_3Br_3$ ✓

(c) $C_2H_4Br_2$ can be refluxed with aqueous sodium hydroxide forming an organic product **B** and bromide ions.

(i) Suggest the formula of the organic product **B**. [1]

$C_2H_4(OH)_2$ ✓

(ii) Describe how would you show that bromide ions had been formed. [3]

Add silver nitrate, ✓ a cream precipitate of silver bromide is formed. ✓ ✗

(d) Briefly describe how ethene can be converted into ethanol using sulphuric acid. [2]

Pass ethene into concentrated sulphuric acid. ✗ ✗

(e) Suggest how ethene can be obtained from ethanol. [2]

Use heat ✓ with a dehydrating agent. ✗

Examiner's Commentary

? For help see Revise AS Study Guide 7.4 and 7.6

(a)(iii) You must give the colour change – orange to colourless.

(b) Many others are possible, e.g. $C_2H_2Br_4$, C_2HBr_5, C_2Br_6.

(c)(ii) You must first add excess nitric acid to remove the sodium hydroxide that would give a precipitate with silver nitrate.

(d) Nothing happens in the cold. To convert ethene to ethanol, we need steam and concentrated sulphuric acid.

(e) You should name the dehydrating agent – excess concentrated sulphuric acid.

GRADE BOOSTER

You must learn the conditions for chemical reactions. Conditions are particularly important for organic reactions where a change of conditions can produce a different final product.

A grade candidate – mark scored 8/10

Ethanoic acid and glucose (molecular formula $C_6H_{12}O_6$) have the same empirical formula.

(a) What is the structural formula of ethanoic acid? [1]

CH_3COOH ✓

(b) Show that ethanoic acid and glucose have the same empirical formula. [2]

Molecular formula of ethanoic acid is $C_2H_4O_2$, simplest formula is CH_2O. ✓

Simplest formula for glucose is CH_2O, so they both have the same empirical formula. ✓

(c) Suggest the product formed when glucose is dehydrated by concentrated sulphuric acid. [1]

carbon ✓

(d) Ethanoic acid can be formed from ethanol.

$$C_2H_5OH(l) + 2[O] \rightarrow CH_3COOH(l) + H_2O(l)$$

(i) State suitable reagents for this reaction and the essential conditions required. [3]

$Cr_2O_7^{2-}$ ✗ and H_2SO_4 ✓ reflux ✓

(ii) State, with reasons, whether or not the formation of ethanoic acid from ethanol is an example of the following types of reaction: [3]

Type of reaction	Reason
Addition	no, two products are formed, not one ✓
Oxidation	yes, oxygen is added ✓
Hydrolysis	yes, water is formed ✗

Examiner's Commentary

Don't forget to include the C in the COOH group when naming acids.

? For help see Revise AS Study Guide 7.5

Six molecules of water would be removed leaving 6C.

You must give the name or formula of the reagents. $Cr_2O_7^{2-}$ is NOT the name or the correct formula of the reagent. Potassium dichromate(VI) ($K_2Cr_2O_7$) is a suitable reagent.

No, hydrolysis is a reaction with water – in this case water is formed.

Exam practice questions

1 An alcohol has a relative molecular mass of 74 and has the following composition by mass: C, 64.9%; H, 13.5%; O, 21.6%.

(a) Calculate the empirical formula of the alcohol and show that its molecular formula is the same as the empirical formula. [4]

(b) Draw the displayed formula of the **four** possible structural isomers of this alcohol. [4]

(c) Compound **F**, one of these isomers, can be oxidised to form a ketone, **G**.

(i) Show the structure of compound **G**. [2]

(ii) Deduce which of the four alcohols in **(b)** is compound **F**. [1]

[Total: 11]

Answers on pages 62–63 Answers on pages 62–63 Answers on pages 62–63

2 **X** is an unsaturated hydrocarbon of relative molecular mass 54. **X** does not have any triple bonds. **X** melts at −109°C and boils at −5°C.

(a) Will **X** be a solid, liquid or gas at room temperature and pressure? Give a reason for your answer. [1]

(b) How would you show that **X** was unsaturated? [2]

(c) At room temperature and pressure, 2.7 g of **X** reacts with 2.4 dm^3 of hydrogen. Show that each molecule of **X** contains two double bonds. [2]

(d) Show that the molecular formula of **X** is C_4H_6. [1]

(e) What is the empirical formula of **X**? [1]

(f) **Y** is an unsaturated hydrocarbon with the molecular formula C_4H_8. Draw possible structures of **two** structural isomers of **Y**. [2]

(g) If you were given a mixture of these two isomers of **Y**, suggest how would you separate them. [1]

(h) Draw the skeletal structure of one of these isomers of **Y**. [1]

(i) Suggest the structure of the polymer formed from one of the isomers of **Y**. (Your diagram should show at least two monomers joined together.) [2]

[Total: 13]

Organic chemistry

Answers on pages 62–63 Answers on pages 62–63 Answers on pages 62–63

3 This question looks at some reactions of halogenoalkanes.

(a) When bromoethane, CH_3CH_2Br is heated with a solution of sodium hydroxide, NaOH, in ethanol, an alkene is formed.

(i) Name the alkene. [1]

(ii) Write an equation for the reaction. [2]

(b) When bromoethane, CH_3CH_2Br, is heated with aqueous sodium hydroxide, NaOH(aq), ethanol is formed.

(i) Outline, using curly arrows, the mechanism of this reaction. [3]

(ii) Name the type of reaction. [2]

(c) The reaction in **(b)** was repeated but using chloroethane, CH_3CH_2Cl, as the halogenoalkane.

Explain why the reactions with bromoethane and chloroethane take place at different rates. [2]

(d) 2-Bromobutane was heated with ethanolic sodium hydroxide.

- A mixture of three isomers **A**, **B** and **C** was formed.
- Isomers **B** and **C** are geometric (*cis–trans*) isomers of one another.
- Each isomer had a molecular formula of C_4H_8.

Draw and name each isomer. [5]

[Total: 15]

Organic chemistry

 Answers on pages 62–63 Answers on pages 62–63 Answers on pages 62–63

4 This question is about two gases **X** and **Y**.

(a) 1 dm^3 of gas **X** at room temperature and pressure (r.t.p.) has a mass of 1.25 g. The empirical formula of **X** is CH_2O. What is the molecular formula of **X**? [2]

(b) Oxidation of **X** gives an acid **Z** with the molecular formula CH_2O_2.

(i) Name acid **Z** and draw its displayed formula. [2]

(ii) How would you show the presence of the OH group in **Z**? [1]

(c) 24 dm^3 of **Y** burns completely in oxygen to give 9.0 g of water, 12 dm^3 of nitrogen. Carbon dioxide is also formed. **Y** has a relative molecular mass of 27. 1 mole of a gas at r.t.p. has a volume of 24 dm^3.

(i) What elements MUST be present in **Y**? [1]

(ii) What is the mass of hydrogen in 9 g of water? [1]

(iii) What is the mass of 12 dm^3 of nitrogen? [1]

(iv) Deduce the formula of **Y**. [1]

(v) Write the equation, including state symbols, for **Y** burning in oxygen. [2]

(d) **Y** reacts slowly with water in the presence of sodium hydroxide to give ammonia and the sodium salt of acid **Z** as the only products.

Write the equation for this reaction. [2]

[Total: 13]

Answers on pages 62–63 Answers on pages 62–63 Answers on pages 62–63

Answers

(1) (a) C : H : O

$\frac{64.9}{12}$ $\frac{13.5}{1}$ $\frac{21.6}{16}$ ✓ = 5.41 : 13.5 : 1.35

Dividing each by 1.35 gives = 4 : 10 : 1 ✓

Empirical formula is $C_4H_{10}O$, which has a mass of 74 ✓ which is the M_r of the molecule. Therefore molecular formula is the same as the empirical formula. ✓

(b)

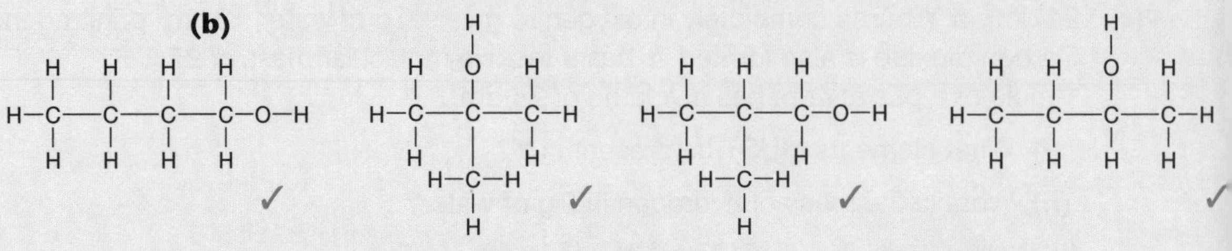

(c) (i)

```
  H  H  O  H
  |  |  ‖  |
H-C--C--C--C-H
  |  |     |
  H  H     H
```

ketone ✓
structure ✓

(ii)

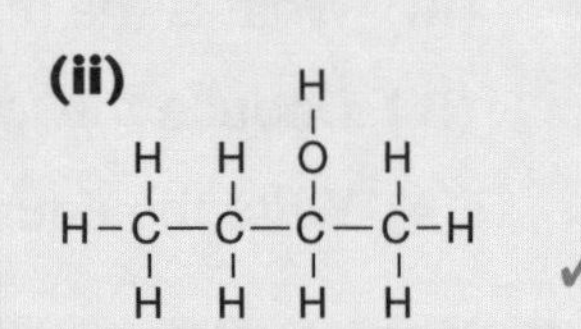

examiner's tip **One mole of hydrogen is required to saturate one double bond.**

(2) (a) gas; m.pt. and b.pt. < room temperature ✓

(b) Add bromine/bromine water; ✓ it changes from orange to colourless. ✓

(c) 54 g reacts with 48 dm^3, ✓ which is 2 moles of hydrogen, therefore it has 2 double bonds. ✓

(d) $4 \times 12 + 6 \times 1 = 54$ ✓

(e) C_2H_3 ✓

(f) two from (2 marks ✓✓)

```
H      CH3     H      H          H3C     CH3        H3C     H
 \    /         \    /              \    /             \    /
  C=C            C=C                 C=C        or      C=C
 /    \         /    \              /    \             /    \
H      CH3     H      CH2-CH3      H      H            H      CH3
```

(g) liquefy and then fractional distillation ✓

examiner's tip **Isomers have different boiling points, and therefore can be separated by this method.**

(h) any one from (one mark)

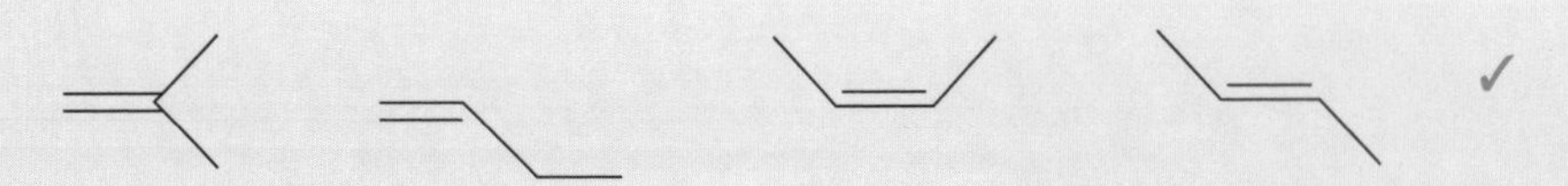

(i) one from

```
 [H   CH3  H   CH3]        [H   CH3  H   CH3]        [H    CH3  H    CH3]
 [|   |    |   |  ]        [|   |    |   |  ]        [|    |    |    |  ]
-[C---C----C---C--]-      -[|   CH2  H   CH2]-      -[C----C----C----C--]-
 [|   |    |   |  ]        [|   |    |   |  ]        [|    |    |    |  ]
 [H   CH3  H   CH3]n       [C---C----C---C--]        [CH3  H    CH3  H  ]n
                           [|   |    |   |  ]
                           [H   H    H   H  ]n
```

(✓ for repeated unit, ✓ for correct structure)

(3) (a) (i) ethene ✓ **(ii)** $CH_3CH_2Br + NaOH \rightarrow C_2H_4 + NaBr + H_2O$ ✓✓

(b) (i)

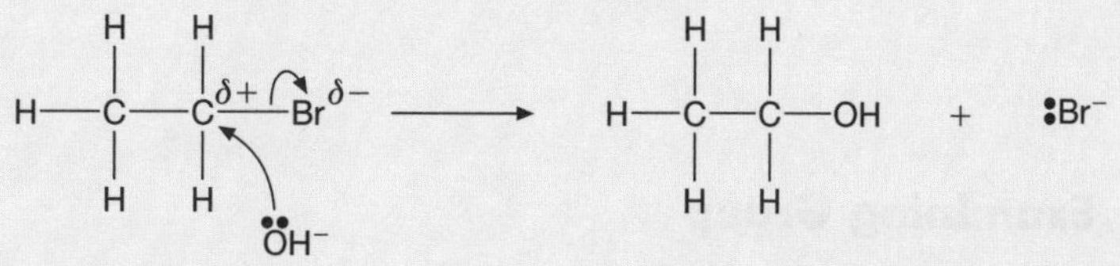

$$H-CH_2-\overset{\delta+}{C}H_2-\overset{\delta-}{Br} \;(\text{attacked by } :OH^-) \longrightarrow H-CH_2-CH_2-OH + :Br^-$$

arrow in from OH^-; ✓ arrow out to Br^-; ✓ dipoles and lone pairs ✓

(ii) nucleophilic ✓ substitution ✓

examiner's tip

Sodium hydroxide dissolved in ethanol removes HBr from bromoethane; when sodium hydroxide is dissolved in water it hydrolyses the bromoethane to ethanol.

(c) The C–Cl bond is stronger than the C–Br bond, ✓ therefore the reaction would take place more slowly. ✓

examiner's tip

The C–Cl bond is more polar than the C–Br bond, therefore the reaction should take place more quickly. It is likely that the reaction will take place more slowly because the breaking of bonds is a more important factor.

(d)

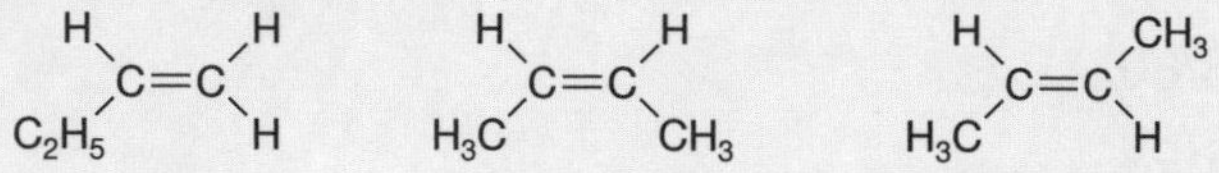

$H(C_2H_5)C=CH_2$	$H(H_3C)C=C(CH_3)H$ (cis)	$H(H_3C)C=C(H)CH_3$ (trans)
but-1-ene	*cis*-but-2-ene	*trans*-but-2-ene

structures ✓✓✓; names: but-1-ene ✓ *cis* and *trans* but-2-ene ✓

---

(4) (a) 24 dm^3 has a mass of 30 g ✓ which is mass of 1 mole of CH_2O = molecular formula ✓

(b) (i) methanoic acid ✓ $H-C(=O)-OH$ ✓

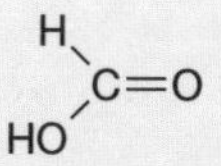

(ii) Add PCl_5 (phosphorus(V) chloride), gives misty fumes of hydrogen chloride. OR add sodium, giving bubbles of hydrogen gas. ✓

(c) (i) carbon, nitrogen and hydrogen ✓

examiner's tip

You cannot tell if Y contained oxygen – it could have come from the oxygen.

(ii) 1 g ✓

(iii) 14 g ✓

(iv) HCN ✓✓

examiner's tip

Since the molecular mass of Y is 27, then 27 − 1 − 14 = 12. So it contains only one carbon atom.

(v) $4HCN(g) + 5O_2(g) \rightarrow 2H_2O(l) + 4CO_2(g) + 2N_2(g)$ equation ✓ balancing ✓

(d) $HCN + H_2O + NaOH \rightarrow HCOONa + NH_3$ left-hand side; ✓ right-hand side ✓

AS Mock Exam 1

Centre number ______________________
Candidate number ______________________
Surname and initials ______________________

Examining Group

Chemistry

Time: 1 hour   Maximum marks: 60

This paper is based on: Chapter 1 Atomic Structure; Chapter 2 Atoms, Moles and Equations; and Chapter 3 Structure and Bonding.

Instructions
Answer **all** questions in the spaces provided. Show all steps in your working.
The marks allocated for each question are shown in brackets.
Any data required for a question are given where appropriate.

Grading
Boundary for A grade 48/60
Boundary for C grade 36/60

---

1 Lithium exists naturally as a mixture of isotopes.

(a) Explain the term *isotopes*.

....

.... [2]

(b) Which isotope is used as the standard against which relative atomic masses are measured?

.... [1]

(c) A sample of lithium has the following percentage composition by mass:
^{6}Li, 7.4%; ^{7}Li, 92.6%

(i) Use this information to help you complete the table below.

isotope	number of	
	protons	neutrons
^{6}Li		
^{7}Li		

[2]

(ii) Calculate the relative atomic mass of this lithium sample. Your answer should be given to three significant figures.

[2]

(d) Relative atomic masses can be determined using a mass spectrometer.
State the main processes that take place in the four regions shown below.

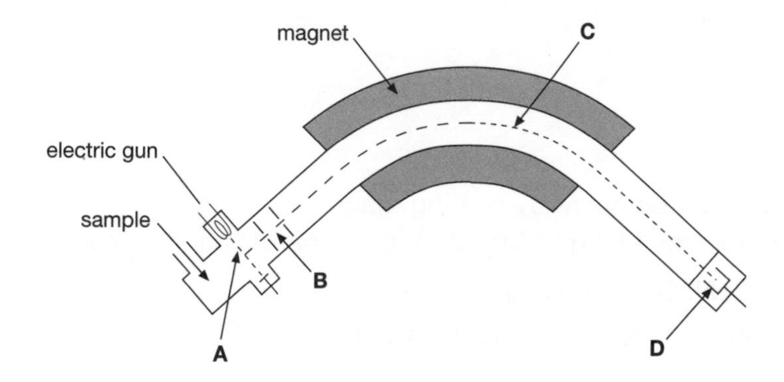

A .... **[1]**

B .... **[1]**

C .... **[1]**

D .... **[1]**

[Total: 11]

2 Magnesium oxide, MgO, is an ionic compound. This question looks at two reactions in which magnesium oxide is a product.

(a) Magnesium oxide can be formed by heating magnesium in oxygen.

(i) Complete the electronic configuration for a magnesium atom.

$1s^2$.. [1]

(ii) Complete the electronic configuration for a magnesium ion.

$1s^2$.. [1]

(iii) Write an equation, including state symbols, for the reaction of magnesium with oxygen.

.... [2]

(iv) Draw a 'dot-and-cross' diagram of magnesium oxide.

[2]

(b) Magnesium oxide can be made by heating magnesium nitrate.
A student decomposed 4.50 g of $Mg(NO_3)_2$ as in the equation below.

(A_r: Mg, 24·3; N, 14·0; O, 16·0)

$2Mg(NO_3)_2(s) \rightarrow 2MgO(s) + 4NO_2(g) + O_2(g)$

(i) Calculate how many moles of $Mg(NO_3)_2$ were heated.

[2]

(ii) Calculate the mass of MgO that was formed.

[2]

(iii) Calculate the total volume of gases formed. Assume that the gas volumes were measured at room temperature and pressure (r.t.p.).
1 mol of gas molecules has a volume of 24.0 dm^3 at r.t.p.

[2]

(c) A sample of magnesium oxide was neutralised by an acid forming a compound **A** with the following percentage composition by mass:

Mg, 21.6%; C, 21.4%; O, 57.0%.

Calculate the empirical formula of **A**.

[2]

[Total: 14]

3 The chemistry of ammonia, NH_3, is influenced by its polarity and its ability to form hydrogen bonds.

(a) Polarity can be explained in terms of electronegativity.

(i) Explain the term *electronegativity*.

....

....

.... [2]

(ii) Why are ammonia molecules polar?

....

.... [1]

(b) The polarity of NH_3 molecules results in the formation of hydrogen bonds.

(i) Draw a diagram to show hydrogen bonding between two molecules of NH_3. Your diagram should include dipoles and lone pairs of electrons.

[3]

(ii) State the H–N–H bond angle in an ammonia molecule.

.... [1]

(c) Ammonia reacts with hydrochloric acid, forming the ammonium ion NH_4^+.

(i) State the H—N—H bond angle in an ammonium ion.

.... [1]

(ii) Explain why the H—N—H bond angle changes during this reaction.

....

....

....

....

.... [3]

[Total: 11]

4 The table below shows some values of melting points and some heat energies needed for melting.

Substance	I_2	NaCl	HF	HCl	HI
Melting point / K	387	1074	190	158	222
Heat energy for melting / kJ mol^{-1}	7.9	28.9	3.9	2.0	2.9

(a) Name three types of intermolecular force.

....

....

.... [3]

(b) **(i)** Describe the bonding in a crystal of iodine.

....

.... [2]

(ii) Name the crystal type which describes an iodine crystal.

.... [1]

(iii) Explain why heat energy is required to melt an iodine crystal.

....

.... [1]

(c) In terms of the intermolecular forces involved, suggest why:

(i) hydrogen fluoride requires more heat energy for melting than does hydrogen chloride

....

....

....

.... [3]

(ii) hydrogen iodide requires more heat energy for melting than does hydrogen chloride.

....

....

.... [2]

(d) (i) Explain why the heat energy required to melt sodium chloride is large.

....

.... [2]

(ii) The heat energy needed to vaporise one mole of sodium chloride (171 kJ mol^{-1}) is much greater than the heat energy required to melt one mole of sodium chloride.
Explain why this is so.

....

.... [1]

(e) In terms of its structure and bonding, suggest why graphite has a very high melting point.

....

....

.... [2]

AQA June 2002

[Total: 17]

5 Well over 2 000 000 tonnes of sulphuric acid, H_2SO_4, are produced in the U.K. each year. This is used in the manufacture of many important materials such as paints, fertilisers, detergents, plastics, dyestuffs and fibres.

The sulphuric acid is prepared from sulphur in a 3 stage process.

Stage 1:

The sulphur is burnt in oxygen to produce sulphur dioxide.

$$S + O_2 \rightarrow SO_2$$

Stage 2:

The sulphur dioxide reacts with more oxygen using a catalyst to form sulphur trioxide.

$$2SO_2 + O_2 \rightarrow 2SO_3$$

Stage 3:

The sulphur trioxide is dissolved in concentrated sulphuric acid to form 'oleum', $H_2S_2O_7$, which is then diluted in water to produce sulphuric acid.

(a) 100 tonnes of sulphur dioxide were reacted with oxygen in stage 2.

Assuming that the reaction was complete, calculate:

(i) how many moles of sulphur dioxide were reacted

M_r: SO_2, 64.1; 1 tonne = 1×10^6 g

[1]

(ii) the mass of sulphur trioxide that formed.

M_r: SO_3, 80.1

[1]

(b) Construct a balanced equation for the formation of sulphuric acid from oleum.

.... [1]

(c) The concentration of the sulphuric acid can be checked by titration. A sample of the sulphuric acid was analysed as follows.

- $10.0\,cm^3$ of sulphuric acid was diluted with water to make $1.00\,dm^3$ of solution.
- The diluted sulphuric acid was then titrated with aqueous sodium hydroxide, NaOH.

$$H_2SO_4(aq) + 2NaOH(aq) \rightarrow Na_2SO_4(aq) + 2H_2O(l)$$

- In the titration, $25.0\,cm^3$ of $0.100\,mol\,dm^{-3}$ aqueous sodium hydroxide required $20.0\,cm^3$ of **diluted** sulphuric acid for neutralisation.

(i) Calculate how many moles of NaOH were used.

[1]

(ii) Calculate the concentration, in $mol\,dm^{-3}$, of the **diluted** sulphuric acid, H_2SO_4.

[2]

(iii) Calculate the concentration, in $mol\,dm^{-3}$, of the original sulphuric acid submitted for analysis.

[1]

OCR Jan 2001

[Total: 7]

AS Mock Exam 2

Centre number ____________________
Candidate number ____________________
Surname and initials ____________________

Examining Group

Chemistry

Time: 1 hour   Maximum marks: 60

This paper is based on: Chapter 4 The Periodic Table and Chapter 5 Chemical Energetics.

Instructions
Answer **all** questions in the spaces provided. Show all steps in your working.
The marks allocated for each question are shown in brackets.
Any data required for a question are given where appropriate.

Grading
Boundary for A grade 48/60
Boundary for C grade 36/60

---

1 The atomic radii of the elements Li to F and Na to Cl are shown in the table below.

element	Li	Be	B	C	N	O	F
atomic radius/nm	0.134	0.125	0.090	0.077	0.075	0.073	0.071
element	Na	Mg	Al	Si	P	S	Cl
atomic radius/nm	0.154	0.145	0.130	0.118	0.110	0.102	0.099

(a) Explain what causes the general **decrease** in atomic radii across each period.

....

....

....

....

.... **[3]**

(b) Explain what causes the general **increase** in atomic radii down each group.

....

....

....

.... [3]

(c) The first ionisation energy of aluminium is $+578\,kJ\,mol^{-1}$.

(i) Define the term *first ionisation energy*.

....

....

....

.... [3]

(ii) Write an equation, with state symbols, for the change that corresponds to the first ionisation energy of aluminium.

.... [2]

(iii) The first ionisation energy of magnesium is $+738\,kJ\,mol^{-1}$.

Explain why, despite having smaller atoms, the first ionisation energy of aluminium is less than that of magnesium.

....

....

.... [2]

[Total: 13]

2 This question concerns elements and compounds from Group 2 of the Periodic Table.

(a) State the trend in reactivity of the Group 2 elements with oxygen.
Explain your answer.

trend in reactivity ....

explanation ....

....

.... **[4]**

(b) The reactions of strontium are typical of a Group 2 element.

Write the formulae for substances **A–D** in the flow chart below.

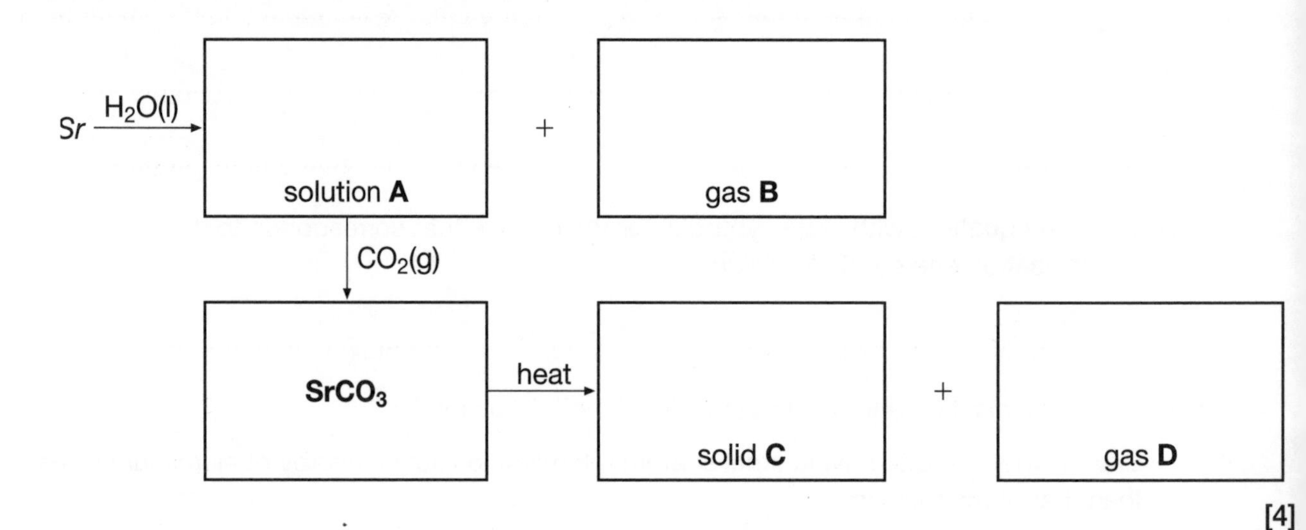

[4]

OCR 2002

[Total: 8]

3 Hydrocarbons, such as heptane C_7H_{16}, are used as fuels making use of their combustion reaction with oxygen. Polluting gases such as NO and NO_2 are also made during combustion in the petrol engines of cars.

(a) Write a balanced equation for the complete combustion of heptane.

.... [2]

(b) **(i)** Explain the term *enthalpy change of formation*.

....

....

.... [2]

(ii) What conditions are needed for this to be a *standard* enthalpy change?

....

.... [2]

(iii) Write an equation to represent the enthalpy change of formation of $NO_2(g)$.

.... [2]

(c) Modern cars have 'catalytic converters' in their exhausts to convert nitrogen oxides into less harmful substances. One reaction that occurs is as follows:

$$4CO + 2NO_2 \rightarrow 4CO_2 + N_2$$

Use the data below to calculate the enthalpy change of this reaction.

compound	ΔH_f/kJ mol^{-1}
NO_2	+33
CO	−111
CO_2	−394

[3]

[Total: 11]

4 This question refers to the hydrogen halides HF and HCl. The table below lists some bond enthalpies which are required in different parts of this question.

bond	average bond enthalpy/kJ mol^{-1}
F—F	+158
Cl—Cl	+244
H—F	+568
H—Cl	+432

(a) Explain the term *bond enthalpy.*

....

....

.... [2]

(b) The hydrogen halides HCl and HF can be made from their elements.
The formation of HCl is exothermic:

$H_2(g) + Cl_2(g) \rightarrow 2HCl(g) \qquad \Delta H = -184\ kJ\ mol^{-1}$

(i) Show that the bond enthalpy of the H—H bond is +436 kJ mol^{-1}.

[2]

(ii) Calculate the enthalpy change for the formation of HF from its elements.

$H_2(g) + F_2(g) \rightarrow 2HF(g)$

[2]

(c) The reaction between hydrogen and chlorine to form hydrogen chloride is exothermic.

(i) Explain why no reaction takes place unless the reactants are sparked.

....

....

.... [2]

(ii) Draw an enthalpy profile diagram to support your answer in **(c)(i)**.

[3]

[Total: 11]

5 Chlorine and its compounds have many uses. Chlorine bleach is used to kill bacteria.

(a) Chlorine bleach is made by the reaction of chlorine with aqueous sodium hydroxide.

$$Cl_2(g) + 2NaOH(aq) \rightarrow NaClO(aq) + NaCl(aq) + H_2O(l)$$

(i) Determine the oxidation number of chlorine in

Cl_2 ....

NaClO ....

NaCl .... **[3]**

(ii) Construct a half equation for the reduction of chlorine in the reaction.

.... **[1]**

(b) Describe a simple test for the presence of a chloride ion. You should write an ionic equation as part of your answer.

test ....

....

equation .... **[3]**

(c) The halogens have different reactivities.

- Explain the trend in reactivity shown by the halogens.
- Describe, with equations and observations, how displacement reactions can be used to show the different reactivities of chlorine, bromine and iodine.

....

....

....

....

....

....

....

....

....

[10]

[Total: 17]

AS Mock Exam 3

Centre number ____________

Candidate number ____________

Surname and initials ____________

Examining Group

Chemistry

Time: 1 hour Maximum marks: 60

This paper is based on: Chapter 6 Rates and Equilibrium and Chapter 7 Organic Chemistry.

Instructions

Answer **all** questions in the spaces provided. Show all steps in your working.
The marks allocated for each question are shown in brackets.
Any data required for a question are given where appropriate.

Grading

Boundary for A grade	48/60
Boundary for C grade	36/60

---

1 Ammonia, NH_3, is manufactured by the Haber process in an exothermic equilibrium reaction.

$$N_2(g) + 3H_2(g) \rightleftharpoons 2NH_3(g) \quad \Delta H = -92\ \text{kJ mol}^{-1}$$

(a) Describe and explain the effect of pressure on the **rate** of this reaction.

....

....

.... [2

(b) State Le Chatelier's principle.

....

....

.... [2

(c) Describe and explain, in terms of Le Chatelier's Principle, how the **equilibrium position** of this reaction is affected by:

(i) increasing the temperature

....

....

.... [2]

(ii) increasing the pressure.

....

....

.... [2]

(d) Why is the temperature used described as a compromise?

....

....

.... [2]

(e) The production of ammonia in the Haber Process is carried out in the presence of an iron catalyst.

State and explain the effect that the catalyst has on:

(i) the rate of ammonia production

....

....

....

....

.... [3]

(ii) the equilibrium position.

....

....

....

....

.... [2]

[Total: 15]

2 The rate of any chemical reaction increases if the temperature is increased.

(a) Draw a diagram to represent the Maxwell–Boltzmann distribution of molecular energies at a temperature T_1 and at a higher temperature T_2.

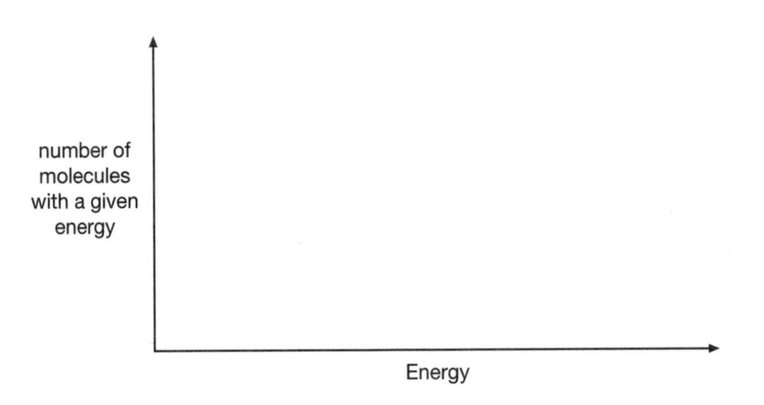

[4]

(b) Use your diagram and the idea of activation energy to explain why the rate of a chemical reaction increases with increasing temperature.

....

....

....

....

....

....

....

.... [3]

Edexcel June 2001

[Total: 7]

3 The hydrocarbons in crude oil can be separated by fractional distillation.

(a) Explain in terms of forces why fractional distillation separates the hydrocarbons in crude oil.

....

....

.... [2]

(b) The alkanes are an example of a homologous series.

(i) Explain the term *homologous series.*

....

....

.... [2]

(ii) What is the general formula for the alkanes?

.... [1]

(iii) What is the formula of the alkane with 15 carbon atoms?

.... [1]

(iv) Calculate the relative molecular mass of pentane.

.... [1]

(c) Undecane, $C_{11}H_{24}$, can be isolated by fractional distillation and then cracked into octane and compound **A**.

(i) Write a balanced equation to represent this cracking of undecane.

.... [1]

(ii) Name compound **A**.

.... [1]

(iii) Explain why the cracking of hydrocarbons is such an important process.

....

....

.... [2]

[Total: 11]

4 Butane, C_4H_{10}, reacts with Cl_2 in the presence of sunlight to form a mixture of chlorinated products, including C_4H_9Cl, formed as shown in the following equation.

$$C_4H_{10} + Cl_2 \rightarrow C_4H_9Cl + HCl$$

(a) Write equations for the following stages in the mechanism of this reaction.

initiation ....

propagation ....

....

termination .... **[4]**

(b) Compound **A** is one of two possible straight-chained structural isomers of C_4H_9Cl. Compound **A** was reacted in the sequence shown below.

A	Stage I OH^-/H_2O reflux →	**B**	Stage II $K_2Cr_2O_7/H_2SO_4$ heat →	**C**	Stage III $K_2Cr_2O_7/H_2SO_4$ reflux →	$CH_3CH_2CH_2COOH$

(i) Identify compounds **A**, **B** and **C**.

compound **A**	compound **B**	compound **C**

[3]

(ii) Explain the role of the hydroxide ion, OH^- in stage I above.

....

.... **[2]**

(iii) For stage II above, state what colour changes take place in the reaction mixture.

from .... *to* .... **[1]**

(iv) Write a balanced equation for the oxidation of **B** to **C**. The oxidising agent can be represented as [O] in your equation.

.... **[1]**

(c) Organic compounds containing halogens have many uses. Compound **A**, a compound in a dry cleaning solvent, has the percentage composition by mass: Cl, 41.6%; C, 14.0%; F, 44.4%. The relative molecular mass of compound **A** is 171.

(i) Calculate the molecular formula of compound **A**.

[3]

(ii) Suggest a possible structure for compound **A**. Name your structure.

name: .... [2]

[Total: 16]

5 Two reactions of but-1-ene, C_4H_8 are shown below.

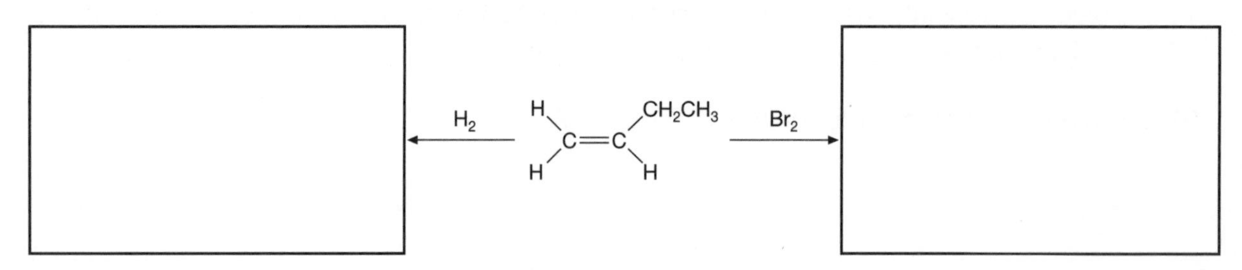

(a) Draw structures for the products of these two reactions in the boxes above. **[2]**

(b) For the reaction between but-1-ene and bromine:

(i) name the type of reaction that takes place

.... **[1]**

(ii) what type of reagent is bromine?

.... **[1]**

(iii) complete the mechanism below.

H
CH_2CH_3
C=C ⟶ ⟶
H
H
Br
|
Br

[3]

(c) But-1-ene reacts with steam in the presence of an acid catalyst to form a mixture of two alcohols. Draw the structure of the two alcohols in the boxes below.

[2]

(d) But-1-ene can be polymerised into poly(but-1-ene).

(i) What type of polymerisation forms poly(but-1-ene)?

.... [1]

(ii) Draw a section of poly(but-1-ene) to show **two** repeat units.

[1]

[Total : 11]

AS Mock Exam 1 Answers

(1) (a) Atoms of the same element with different numbers of neutrons ✓ and different masses. ✓ [2]

(b) carbon-12 ✓ [1]

(c) **(i)** ^{6}Li: $3p^+$, 3n; ✓ ^{7}Li: $3p^+$, 4n ✓ [2]

(ii) $8 \times 6/100 + 92 \times 7/100 = 6.926$ ✓ $= 6.93$ (to 3 sig figs) ✓ [2]

(d) A: ionisation; ✓ B: acceleration; ✓ C: deflection; ✓ D: detection ✓ [4]

[Total: 11]

(2) (a) **(i)** $1s^2 2s^2 2p^6 3s^2$ ✓ [1]

(ii) $1s^2 2s^2 2p^6$ ✓ [1]

(iii) $2Mg(s) + O_2(g) \rightarrow 2MgO(s)$ equation: ✓ state symbols: ✓ [2]

(iv) dot and cross ✓; charges ✓ [2]

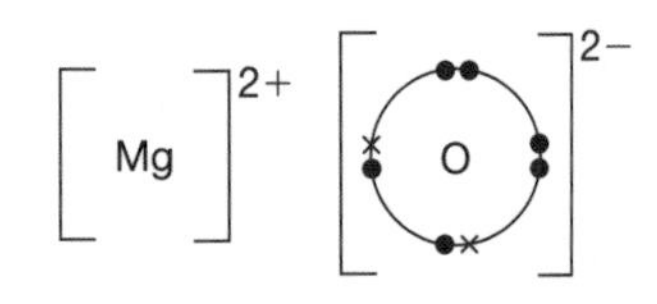

(b) **(i)** molar mass of $Mg(NO_3)_2 = 148.3\ g\ mol^{-1}$ ✓
moles $Mg(NO_3)_2 = 4.50/148.3 = 3.03 \times 10^{-2}$ mol ✓ [2]

(ii) molar mass of $MgO = 40.3\ g\ mol^{-1}$ ✓
mass $MgO = 3.03 \times 10^{-2} \times 40.3 = 1.22$ g ✓ [2]

(iii) moles gas molecules $= 5/2 \times 3.03 \times 10^{-2}$ ✓
volume gas $= 24.0 \times 5/2 \times 3.03 \times 10^{-2} = 1.82\ dm^3$ ✓ [2]

(c) ratio $Mg : C : O = 21.6/24.3 : 21.4/12 : 57.0/16 = 0.89 : 1.78 : 3.56$ ✓
empirical formula $= MgC_2O_4$ ✓ [2]

[Total: 14]

(3) (a) **(i)** Attraction (of an atom) for electrons ✓ in a (covalent) bond ✓ [2]

(ii) N and H have different electronegativities ✓ [1]

(b) **(i)** dipole; ✓ H bond; ✓ involvement of lone pair ✓

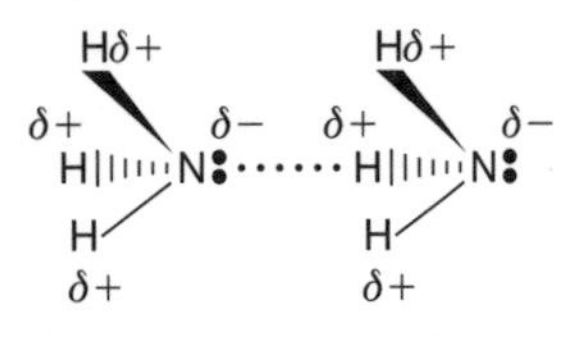

[3]

(ii) 107° ✓ [1]

(c) **(i)** 109.5° ✓ [1]

(ii) NH_3 has 3 bonded pairs and a lone pair of electrons ✓
NH_4^+ has 4 bonded pairs; ✓ lone pair has a greater repelling effect than bonding pair ✓ [3]

[Total: 11]

(4) (a) van der Waals ✓
dipole–dipole ✓
hydrogen bonding ✓ [3]

(b) (i) covalent between atoms ✓
van der Waals between molecules ✓ [2]

(ii) simple molecular ✓ [1]

(iii) van der Waals forces are broken ✓ [1]

(c) (i) HF has hydrogen bonding ✓
HCl has dipole–dipole forces ✓
hydrogen bonding is stronger than dipole–dipole forces ✓ [3]

(ii) HI has more electrons than HCl ✓
intermolecular forces in HI are stronger than those in HCl ✓ [2]

(d) (i) NaCl is ionic ✓
NaCl has strong forces between ions ✓ [2]

(ii) all the bonds must be broken ✓ [1]

(e) graphite is giant molecular ✓
graphite has strong covalent bonds ✓ [2]

[Total: 17]

(5) (a) (i) moles $SO_2 = \frac{100 \times 10^6}{64.1} = 1.56 \times 10^6$ ✓ [1]

(ii) mass $SO_3 = 1.56 \times 10^6 \times 80.1 = 125 \times 10^6$ g or 125 tonne ✓ [1]

(b) $H_2S_2O_7 + H_2O \rightarrow 2H_2SO_4$ ✓ [1]

(c) (i) $0.100 \times 25/1000 = 2.5 \times 10^{-3}$ mol ✓ [1]

(ii) moles $H_2SO_4 = 1.25 \times 10^{-3}$ ✓
concentration $H_2SO_4 = 1.25 \times 10^{-3} \times 1000/20 = 0.0625$ mol dm^{-3} ✓ [2]

(iii) $100 \times 0.0625 = 6.25$ mol dm^{-3} ✓ [1]

[Total: 7]

AS Mock Exam 2 Answers

(1) (a) The nuclear charge increases ✓ as electrons are added to same shell. ✓ The electrons experience greater attraction, reducing atomic radius. ✓ [3]

(b) The number of shells increases. ✓ Electronic shielding increases. ✓ Electrons experience less attraction, increasing atomic radius. ✓ [3]

(c) (i) The energy change when each atom in 1 mole ✓ of gaseous atoms ✓ loses an electron ✓ (to form 1 mole of gaseous 1+ ions). [3]

(ii) $Al(g) \rightarrow Al^+(g) + e^-$ equation; ✓ state symbols ✓ [2]

(iii) In Al, highest energy electron is in 3p orbital. In Mg, highest energy electron is in 3s orbital. ✓ 3p is higher energy than 3s. ✓ [2]

[Total: 13]

(2) (a) *trend in reactivity*: more reactive down group ✓
explanation: electrons lost more easily/ionisation energies decrease/less attraction or pull ✓
some attempt to relate this increase in size of atom / more shells / energy levels ✓
and increase in shielding ✓ [4]

(b)

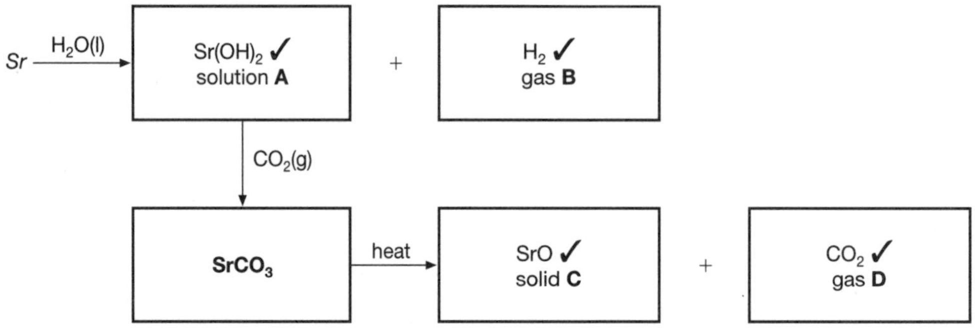

Sr —$H_2O(l)$→ [$Sr(OH)_2$ ✓ solution **A**] + [H_2 ✓ gas **B**]

[$Sr(OH)_2$ solution **A**] —$CO_2(g)$→ [$SrCO_3$]

[$SrCO_3$] —heat→ [SrO ✓ solid **C**] + [CO_2 ✓ gas **D**]

[4]

[Total: 8]

(3) (a) $C_7H_{16}(l) + 11O_2(g) \rightarrow 7CO_2(g) + 8H_2O(l)$ products ✓ balanced ✓ [2]

(b) (i) The enthalpy change that accompanies the formation of 1 mole of a compound ✓ from its constituent elements. ✓ [2]

(ii) 100 kPa ✓ and a stated temperature, usually 298 K ✓ [2]

(iii) $\frac{1}{2}N_2(g) + O_2(g) \rightarrow NO_2(g)$
reactants and products ✓ forming 1 mol NO_2 ✓ [2]

(c) $\Delta H = 4 \times -394 - (4 \times -111 + 2 \times +33)$ ✓ ✓ (2nd mark for correct cycle)
$= -1198 \text{ kJ mol}^{-1}$ ✓ [3]

[Total: 11]

(4) (a) Bond enthalpy is the enthalpy change required to break ✓ and separate 1 mole of bonds ✓ in the molecules of a gaseous element or compound so that the resulting gaseous species exert no forces upon each other. **[2]**

(b) (i) bond enthalpy(H—H) $= -184 - 244 + (2 \times 432)$ ✓ $= +436\ \text{kJ mol}^{-1}$ ✓ **[2]**

(ii) $\Delta H(\text{H—H}) = 436 + 158 - (2 \times 568)$ ✓ $= -542\ \text{kJ mol}^{-1}$ ✓ **[2]**

(c) (i) Activation energy has to be overcome before a reaction can take place. ✓
At room temperature, insufficient molecules exceed the activation energy. ✓ **[2]**

(ii)

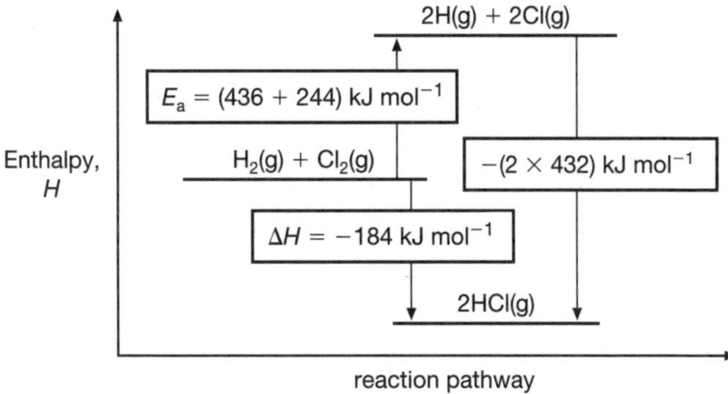

E_a shown; ✓ bonds forming shown; ✓ correct ΔH shown ✓ **[3]**

[Total: 11]

(5) (a) (i) Cl_2: 0; ✓ NaClO: +1; ✓ NaCl: −1 ✓ **[3]**

(ii) $\frac{1}{2}Cl_2 + e^- \rightarrow Cl^-$ ✓ **[1]**

(b) *test:* Add $AgNO_3$(aq). ✓ White precipitate forms. ✓
equation: $Ag^+(aq) + Cl^-(aq) \rightarrow AgCl(s)$ ✓ **[3]**

(c) Reactivity decreases down group/ $Cl_2 > Br_2 > I_2$ ✓

As group descends,

more shells are added / increasing radius of atom ✓
and increasing electron shielding ✓
at the edge of the atom, there is less attraction from the nucleus ✓
electrons gained less easily ✓

with chlorine,

Cl_2 + bromide → orange/brown/yellow/red in organic solvent ✓
$Cl_2 + 2Br^- \rightarrow Br_2 + 2Cl^-$ / $Cl_2 + 2NaBr \rightarrow Br_2 + 2NaCl$ ✓
$Cl_2 + 2I^- \rightarrow I_2 + 2Cl^-$ / $Cl_2 + 2NaI \rightarrow I_2 + 2NaCl$ ✓

with bromine,

Br_2 + iodide → brown/purple with organic solvent ✓
$Br_2 + 2I^- \rightarrow I_2 + 2Br^-$ / $Br_2 + 2NaI \rightarrow I_2 + 2NaBr$ ✓ **[10]**

[Total: 17]

AS Mock Exam 3 Answers

(1) (a) Raising the pressure increases the concentration of gas molecules ✓ causing an increased rate of collision and faster reaction. ✓ **[2]**

(b) A change in conditions shifts the equilibrium position ✓ in the direction that minimises the effect of the change. ✓ **[2]**

(c) (i) To relieve the effect of increased temperature, the equilibrium moves to the left ✓ in the endothermic direction. ✓ **[2]**

(ii) To relieve the effect of increased pressure, the equilibrium moves to the right ✓ in the direction that produces less gas molecules. ✓ **[2]**

(d) The temperature used is sufficiently high to allow the reaction to occur at a realistic rate, ✓ but not too high to give a minimal equilibrium yield of ammonia (equilibrium yield decreases with increasing temperature). ✓ **[2]**

(e) (i) A catalyst speeds up the reaction rate without itself changing, ✓ by providing an alternative route ✓ with a lower activation energy. ✓ **[3]**

(ii) The equilibrium position is unchanged. ✓
The rates of forward and reverse reactions increase by the same amount. ✓ **[2]**

[Total: 15]

(2) (a)

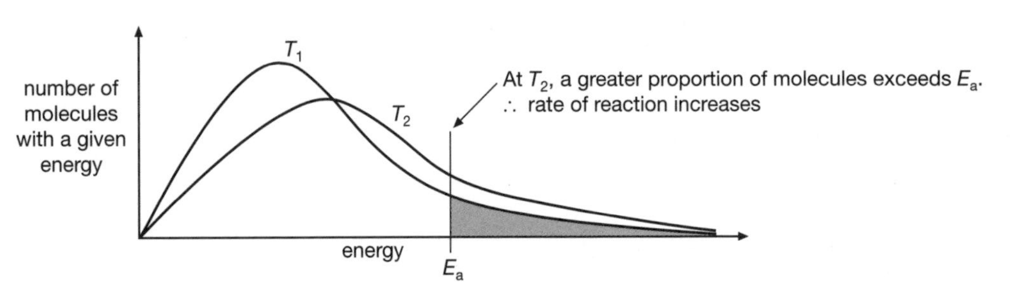

Correct shape drawn for 1 curve, starting at 0 and not touching x-axis ✓
2nd curve correctly drawn for T_2 below level of T_1 ✓ and displaced to the right ✓
At a higher temperature, a larger proportion of molecules exceed the activation energy of the reaction. ✓ **[4]**

(b) E_a labelled on diagram across both curves ✓
For a reaction, those molecules which exceed E_a may react. ✓
At T_2, a greater proportion of molecules exceeds E_a.
∴ more effective collisions ✓

[3]

[Total: 7]

(3) (a) As carbon chain length increases, boiling point increases ✓ because van der Waals' forces increase between molecules. ✓ **[2]**

(b) (i) Each successive member differs by $-CH_2-$ and has the same general formula. ✓
The members of a homologous series have the same functional group and react similarly. ✓ **[2]**

(ii) C_nH_{2n+2} ✓ **[1]**

(iii) $C_{15}H_{32}$ ✓ **[1]**

(iv) 72 ✓ **[1]**

(c) (i) $C_{11}H_{24} \rightarrow C_8H_{18} + C_3H_6$ ✓ **[1]**

(ii) propene ✓ **[1]**

(iii) alkenes are needed to make polymers. ✓ Cracking converts hydrocarbons in low demand into shorter chain alkanes in greater demand. ✓ **[2]**

[Total: 11]

(4) (a) $Cl_2 \rightarrow 2Cl\bullet$ ✓ **[1]**
$C_4H_{10} + Cl\bullet \rightarrow C_4H_9\bullet + HCl$ ✓ **[1]**
$C_4H_9\bullet + Cl_2 \rightarrow C_4H_9Cl + Cl\bullet$ ✓ **[1]**
$2C_4H_9\bullet \rightarrow C_8H_{18}$ or $2Cl\bullet \rightarrow Cl_2$ or $C_4H_9\bullet + Cl\bullet \rightarrow C_4H_9Cl$ ✓ **[1]**

(b) (i) A: $CH_3CH_2CH_2CH_2Cl$; ✓ B: $CH_3CH_2CH_2CH_2OH$; ✓ C: $CH_3CH_2CH_2CHO$ ✓ **[3]**

(ii) OH^- behaves as a nucleophile ✓ by donating an electron pair ✓ **[2]**

(iii) from orange to green ✓ **[1]**

(iv) $CH_3CH_2CH_2CH_2OH + [O] \rightarrow CH_3CH_2CH_2CHO + H_2O$ ✓ **[1]**

(c) (i) molar ratio: Cl 41.6/35.5 : C 14.0/12 : F 44.4/19 ✓
empirical formula = $ClCF_2$ ✓
molecular formula = $Cl_2C_2F_4$ ✓ **[3]**

(ii)

```
    F   F                  F   F
    |   |                  |   |
Cl—C—C—Cl      or      F—C—C—Cl      ✓
    |   |                  |   |
    F   F                  F   Cl
```

1,2-dichloro-1,1,2,2-tetrafluoroethane or 1,1-dichloro-1,2,2,2-tetrafluoroethane ✓ **[2]**

[Total: 16]

(5) (a)

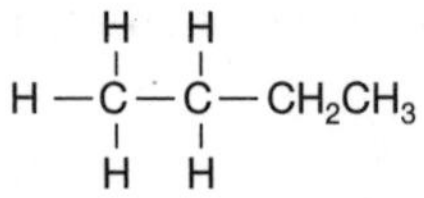

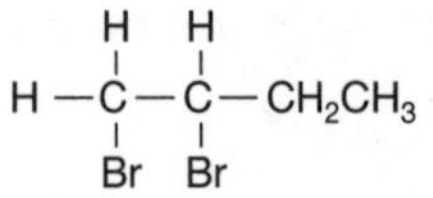

$$H-\overset{H}{\underset{H}{C}}-\overset{H}{\underset{H}{C}}-CH_2CH_3$$ ✓

$$H-\overset{H}{\underset{Br}{C}}-\overset{H}{\underset{Br}{C}}-CH_2CH_3$$ ✓

[2]

(b) (i) addition ✓ **[1]**

(ii) electrophile ✓ **[1]**

(iii)

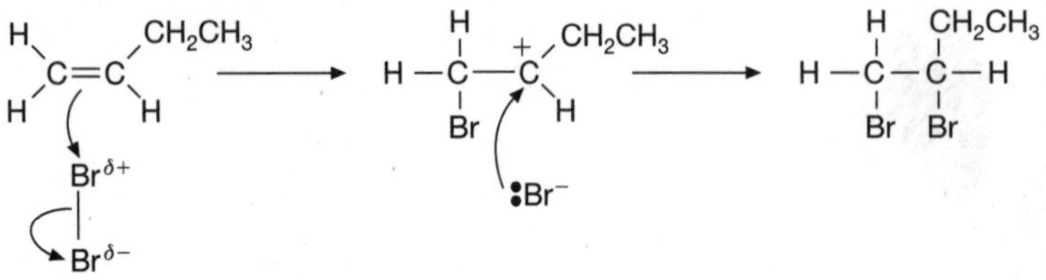

$$H_2C{=}CH(CH_2CH_3) + Br^{\delta+}{-}Br^{\delta-} \longrightarrow H-\overset{H}{\underset{Br}{C}}-\overset{+}{C}H(CH_2CH_3) + {:}Br^- \longrightarrow H-\overset{H}{\underset{Br}{C}}-\overset{CH_2CH_3}{\underset{Br}{C}}-H$$

dipole or arrow from double bond; ✓ carbonium ion; ✓ arrow from Br^- ✓ **[3]**

(c)

$CH_3-CH_2-CH_2-CH_2-OH$ ✓

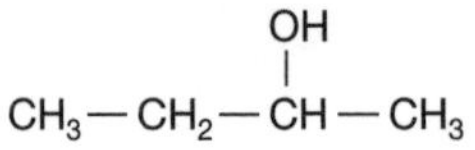

$CH_3-CH_2-\overset{OH}{CH}-CH_3$ ✓

[2]

(d) (i) addition ✓ **[1]**

(ii)

$$-\overset{H}{\underset{H}{C}}-\overset{C_2H_5}{\underset{H}{C}}-\overset{H}{\underset{H}{C}}-\overset{C_2H_5}{\underset{H}{C}}-$$ ✓

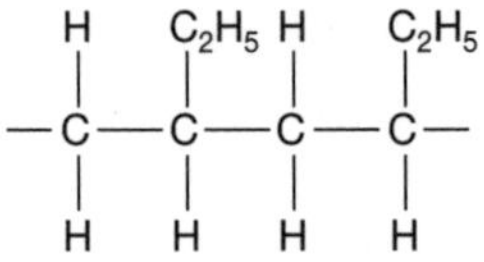

[1]

[Total : 11]